ROUTLEDGE LIBRARY E
ENERGY ECONOM

Volume 23

OPTIMAL PRICING
AND INVESTMENT IN
ELECTRICITY SUPPLY

OPTIMAL PRICING AND INVESTMENT IN ELECTRICITY SUPPLY

An Essay in Applied Welfare Economics

RALPH TURVEY

Routledge
Taylor & Francis Group

LONDON AND NEW YORK

First published in 1968 by George Allen and Unwin Ltd

This edition first published in 2018
by Routledge
2 Park Square, Milton Park, Abingdon, Oxon OX14 4RN

and by Routledge
711 Third Avenue, New York, NY 10017

Routledge is an imprint of the Taylor & Francis Group, an informa business

British Library Cataloguing in Publication Data
A catalogue record for this book is available from the British Library

ISBN: 978-1-138-10476-1 (Set)
ISBN: 978-1-315-14526-6 (Set) (ebk)
ISBN: 978-1-138-50115-7 (Volume 23) (hbk)
ISBN: 978-1-138-50126-3 (Volume 23) (pbk)
ISBN: 978-1-315-14420-7 (Volume 23) (ebk)

Publisher's Note
The publisher has gone to great lengths to ensure the quality of this reprint but points out that some imperfections in the original copies may be apparent.

Disclaimer
The publisher has made every effort to trace copyright holders and would welcome correspondence from those they have been unable to trace.

OPTIMAL PRICING AND INVESTMENT IN ELECTRICITY SUPPLY

An Essay in Applied Welfare Economics

BY

RALPH TURVEY

London

GEORGE ALLEN AND UNWIN LTD

RUSKIN HOUSE   MUSEUM STREET

Printed in Great Britain
in 10 on 12-point Times type
by Alden & Mowbray Ltd
at the Alden Press, Oxford

TO RONNIE

PREFACE

This book is about the economics of electricity supply in general, not about the particular problems of England and Wales. The only limitation on its generality is that the special problems of hydro-electric power are not discussed.

It is aimed both at economists and at those engineers and administrators who are concerned with electricity supply. For the economists it might be described as an essay in applied welfare economics. For those who are not economists by training it might be called an analysis of the 'background philosophy' of investment criteria and tariff policy.

There is a small number of electrical engineers who have specialized in system planning scattered among the world's major electricity undertakings, consulting firms and electrical manufacturers. These experts, who have developed optimization techniques to an impressive extent unrealized by most economists, may find the earlier chapters of this book an inadequately simple guide to investment planning. But it is not intended to explain how to do this work; the aim is to set out its characteristics and to show its implications for cost analysis. It is rarely understood that the dual of cost minimization is the structure of marginal cost; in too many undertakings the commercial side of the work is divorced from the planning side intellectually as well as organizationally.

A very large number of people have helped me to learn about the economics of electricity and I cannot possibly mention them all. Those who have made substantial comments on drafts of this book include J. C. J. Allen, T. W. Berrie, J. K. Delson, R. Janin (E.D.F.), P. T. McIntosh, R. W. Orson, A. G. Oughton, N. D. A. Pallister, R. Rees, C. S. Sinnott, P. E. Watts and I. J. Whitting. This is such an impressive list that I can hardly make the customary disclaimer that I alone am responsible for any technical mistakes that remain!

Finally, I must thank Sir Ronald Edwards, to whom this book is dedicated. He enabled me to spend three years in a fascinating industry with congenial colleagues by appointing me Chief Economist to the Electricity Council. The views expressed in this book being entirely my own, neither he nor any other member of the Council necessarily agrees with them, though I hope that if they do not do so now they will in the future. The book was finished before I became a Member of the National Board for Prices and Incomes.

London, September 1967

CONTENTS

PREFACE   *page* IX

1. Minimum Cost Operation of a Thermal Power System   1
2. The Optimal Plant Mix in the Plant Programme   12
3. The Background Plan   27
4. Long-Run Marginal Cost   44
5. Risk and Uncertainty in Generation   60
6. Investment and Cost Structure in Distribution   68
7. The Quality of Service   82
8. Marginal Costs and Optimal Pricing   86
9. Tariffs   94
APPENDIX: The Economic Analysis of Choice of
            Transformer   108
INDEX   123

xi

1

MINIMUM COST OPERATION
OF A THERMAL POWER SYSTEM

STATEMENT OF THE PROBLEM

At any moment of time, a power system can be regarded as consisting of a number of nodes, all of which are linked to some of the others by transmission lines. Each node contains generating capacity and/or a load to be met. Coal can be supplied from a number of mines and transported to any of the coal-burning generating stations; fuel for oil, gas and nuclear generating stations is assumed to be supplied c.i.f. at the stations in question.

The load to be met at any node is the load of a distribution network, so that such a node is a point where transformers reduce voltage to distribution voltage and the current is fed into the distribution network supplied by that node. Thus the load is the sum of the loads of the consumers served by that network plus its distribution losses.

The short-run problem considered in this chapter is to meet the given loads at standard voltage and frequency at minimum total cost. This reduces to minimizing the sum of fuel purchase, fuel transport and other running costs, for capacity is here taken as given and all other costs can be supposed to depend upon capacity. It will be assumed in what follows that this is in fact the case and other running costs will be neglected in order to make the analysis simple.

Two facets of the problem can be distinguished. One, which has a time-scale of up to a year, involves the pattern of fuel purchases and flows, the general level of operation of each generating station and the periods when plant is withdrawn from service for planned maintenance. The other is the hour to hour operation of the system, sometimes known as 'despatching', i.e. determining which stations generate and at what rate hour by hour. (In the British electricity supply system the unit of time is the half-hour, but in this book it will be taken to be the hour, simply to simplify exposition.) These two facets are discussed in turn in the following pages. Throughout, attention is concentrated on the basic economic aspects of the problem and it must be clearly understood that the discussion pays no attention at all to all sorts of practical engineering aspects which necessarily loom large in the eyes of the engineers whose job it is to face the problem.

MARGINAL GENERATING COST*

A generating station usually contains several sets or units, i.e. in the case of natural gas, oil or coal-fired stations, combinations of boiler, turbine and auxiliary plant. Nowadays, these sets are usually designed so as to function independently of one another. The marginal cost per kWh sent out by any set in such a station is (i) the marginal cost of heat delivered per BTU divided by (ii) marginal kWh sent out per BTU.

(i)   The marginal cost of heat delivered depends upon the cost of fuel, including transport and handling, and its calorific value.

(ii)   Marginal kWh sent out per BTU depends upon four main factors:

— boiler efficiency (which may vary with output)
— the turbine incremental heat rate. (This is constant with throttle-governing plant and rises with output in the case of multi-valve turbines.)
— units (kWh) sent out are some 5% less than units generated (at full load) because of the power consumed by such auxiliaries as water pumps, fans, coal-pulverizing equipment
— performance factor which, to quote Mr Scott, 'covers the somewhat intangible factors of age, design limitations, and generally what may be described as the peculiarities of the station. It can only be assessed from actual operating experience'.

In the following discussion it will be supposed that both (i) and (ii) are constant with respect to output so that the marginal cost curve

---

* 'Marginal cost' is the first derivative of cost with respect to output. This section and the later section on despatching are largely based on E. C. Scott: 'Power System Operation: II, Economic Operation (Steam Plant)', *Electrical Journal*, March 10, 1961; and E. C. Scott *et al.*: 'Criteria for Economic Load Despatching on the British Grid System', C.I.G.R.E. 1964, paper no. 326.

of each set is horizontal up to capacity output. This is reasonably close to the truth for British conditions. When it is not true, as in many U.S. electric utilities, the problem of hour to hour optimization, known as 'economic despatching' is much more complicated than the following discussion suggests. But while the extra complications are of central importance to the operations engineer (the despatcher) they are irrelevant in the context of this book. The minimum cost solution, however complex, still yields a determinate marginal generating cost under given and normal conditions. This is all that is required later on. To allow for nuclear stations, the marginal cost of generation is simply taken as given, such problems of nuclear fuel costs as the techniques of stopping and starting and the valuation of plutonium not being analysed here.

OPTIMISING FUEL CONSUMPTION

We can now examine the nature of the optimization problem for a period stretching several months ahead. As already stated, this involves determining fuel flows, planned outages for maintenance and (in broad outline only) the operating regime for each set in the system so as to minimize the sum of fuel and transport costs. The analysis applies to an ideal situation where coal from each mine is priced at its marginal cost (or the electricity industry owns the coal mines) where coal is transported at marginal cost and where natural gas and oil are delivered to power stations at marginal cost. But the analysis applies equally well to a non-ideal situation where fuels are not made available at marginal cost. In this case the electricity industry can act on the basis of the marginal cost to itself of the different fuels and seek to minimize its own money costs.

The relevant factors are as follows:

(*a*) The load to be met at each node. In principle, what is required is knowledge of the load hour by hour, but the impossibility of such detailed forecasting and computational difficulties both require simplification. We therefore suppose that the period under study is divided into sub-periods (e.g. night-hours, weekday mornings and so on) and that the average load for each sub-period is specified for each node.

(*b*) The sent-out capacity, the kWh sent out per BTU and the required minimum outage period for maintenance of each set.

3

(c) The marginal cost per BTU of coal at each mine, the maximum output of each mine and transport cost from each mine to each coal-burning station.

(d) The marginal cost of oil or natural gas at each oil- or natural gas-burning station and the marginal generating cost of each nuclear station.

(e) The transmission capacity between each pair of nodes and the transmission losses involved.

Given all these factors, linear programming is one of several possible methods which can be used to determine the optimum either from a national point of view (using social marginal costs) or from the point of view of the electricity industry alone (using marginal costs to that industry). These two optima will coincide only if all prices are equal to social marginal costs. The programming solution can only be an approximation because certain considerations will have to be ignored or simplified. Thus there is an implicit assumption that coal flows take place at a uniform rate throughout the period, whereas generation by coal-burning stations will vary over time. Consequently any limitation on coal stocking or destocking at the coal-burning stations is ignored. Then there is the point that a complete analysis of transmission losses (which depend upon the whole pattern of load flow) would be excessively complicated and non-linear. Finally, the number of sub-periods required to deal adequately with the problem of planning plant outages for maintenance may be much larger than the number required for the rest of the analysis.

However such problems are dealt with, the broad outlines of a linear programming formulation of the optimization problem are clear enough. Fuel and transport costs are to be minimized subject to such constraints as:

— the loads sent out by all sets must be sufficient to meet the required loads plus transmission losses in all sub-periods
— the load sent out by each set in each sub-period must not exceed its capacity
— the coal delivered from each mine must not exceed its output capacity
— the fuel consumed by each set must be sufficient to generate its sent-out load over the whole period
— the load on each transmission line must not exceed its capacity in any sub-period

4

— the times when each set is not generating (and each transmission line is not used) must include consecutive periods longer than or equal to its required minimum outage times for maintenance jobs.

The nature of an optimal solution can be shown diagrammatically for the special case of a system of only two nodes in a single sub-period, where both nodes contain both a load and coal-burning generating sets.

In the following diagram the required loads in each of the two nodes are shown as the point labelled 'Load requirement' in the sw quadrant. The combinations of generation in each node which will meet this requirement are shown by the heavy curve passing through this point. Any excess of generation over load is transmitted to the node and this involves some transmission loss. The heavy curve is bounded at its upper end by the generating capacity constraint in node *i* (marked on the w axis) and at its lower end by the trans-mission constraint, so that the generating capacity constraint in node *x* (marked on the s axis) is not relevant.

The thermal efficiency curves in the NW and SE quadrants trans-form kWh sent out into required coal inputs at each node. Hence given the 'Alternative generation requirements' curve in the sw quadrant, the 'Alternative coal requirements curve' can be derived in the NE quadrant. Two isocost lines are shown in this quadrant too. These each show combinations of coal supplies delivered to the two nodes which all have the same total fuel and transport cost. Tangency with the Alternative coal requirements curve gives the optimal solu-tion shown by the dashed line going right round the diagram. It shows node *x* generating more than it consumes and exporting to node *i*.

The diagram has been drawn with non-linearity (transmission losses) and with load requirements such that none of the constraints on mine, generating and transmission capacities is operative. The solution shown is thus not a corner solution and its characteristics can therefore be stated in terms of the satisfaction of certain marginal, i.e. first-order, conditions.

The slope of the Alternative coal requirements curve can be seen to depend upon the slope of the Alternative generation requirements

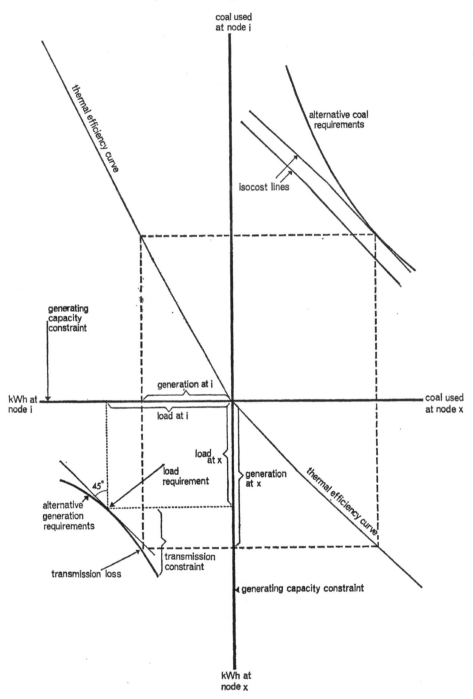

curve and the slopes of the two thermal efficiency curves. Thus it can be written:

$$\frac{(1-\alpha)t^i}{t^x}$$

where $\alpha =$ incremental transmission losses (the supply of $(1-\alpha)$ extra kWh at i from x requires the generation of 1 extra kWh at x)

$t^i =$ incremental thermal efficiency at i, BTU per kWh

$t^x =$ incremental thermal efficiency at x, BTU per kWh

The slope of an isocost line, on the other hand, is the ratio:

$$\frac{c^x}{c^i}$$

where c represents the marginal cost of coal per BTU. Hence the tangency of the two curves in the NE quadrant implies:

$$\frac{(1-\alpha)t^i}{t^x} = \frac{c^x}{c^i}$$

which gives:

$$(1-\alpha)c^i t^i = c^x t^x$$

Now ct is simply the marginal cost of a kWh sent out. Writing m for this marginal cost of generation, we therefore have the intuitively obvious result:

$$m^i = \frac{m^x}{1-\alpha}$$

i.e. the incremental cost of increasing the load by 1 kWh at node i must be the same whether the extra unit is generated in i or is provided by generating and transmitting more from x.* More generally,

---

* A proof for n transmission lines has been set out as follows by N. D. A. Pallister:

Let there be m nodes $N_1, N_2, \ldots N_m$.

Let the power transfers from N_i to N_j be T_{ij} at the middle of the line. If no line exists between N_i and N_j then $T_{ij} = 0$.

Let the power generated at N_i be M_i.

Let the demand at N_i be D_i.

Let the marginal generating cost at N_i at generating capacity M be $g_i(M)$.

Let the power lost between N_i and N_j be $\lambda_{ij} T_{ij}^2$, and assume that the loss rate is linear.

By definition $\lambda_{ij} = \lambda_{ji}$.

[*continued on next page.*

7

in the absence of generating and transmission constraints, optimization requires that the difference in marginal generating cost between any pair of nodes does not exceed α times the lower of the two marginal costs where α is the incremental transmission loss between them. Electrical engineers usually express this in the form:

$$\frac{\mathrm{d}F_n}{\mathrm{d}P_n} + \lambda \frac{\partial P_L}{\partial P_n} = \lambda$$

where F_n is generating cost at the nth set, P_n is its output and P_L is system transmission losses.

The operative constraints are:

(i) the total power generated is equal to the total demand plus losses in transmission

(ii) the power generated at any one node is equal to the demand at that node plus the exported power from the node

(iii) the power flowing in the middle of a line from N_i to N_j is minus the power flowing at the same point from N_j to N_i.

The total generating cost of providing generation of M_i at N_i is $\int_0^{M_i} g_i (M)\mathrm{d}M$.

Using the Lagrange Multiplier Theory the problem becomes one of minimizing the total generating cost subject to the constraints mentioned above.

$$\text{Let } S = \sum_{i=1}^{m} \int_0^{M_i} g_i(M)\mathrm{d}M$$

$$+ \ \mu \left\{ \sum_{i=1}^{m}(M_i - D_i) - \tfrac{1}{2} \sum_{i=1}^{m} \sum_{j=1}^{m} \lambda_{ij} T_{ij}^2 \right\}$$

$$+ \ \sum_{i=1}^{m} \mu_i \left\{ (M_i - D_i) - \sum_{j=1}^{m} T_{ij} - \tfrac{1}{2} \sum_{j=1}^{m} \lambda_{ij} T_{ij}^2 \right\}$$

The optimum is obtained by equating the partial differentials of S with respect to M_i and T_{ij} (the unknowns) to zero and applying the constraints.

$$\frac{\partial S}{\partial M_i} = g_i(M_i) + \mu + \mu_i = 0 \tag{1}$$

$$\frac{\partial S}{\partial T_{ij}} = -\mu\lambda_{ij}T_{ij} - \mu\lambda_{ij}T_{ij} - \mu_i + \mu_j - \mu_i\lambda_{ij}T_{ij} - \mu_j\lambda_{ij}T_{ij} = 0 \tag{2}$$

Eliminating μ_i and μ_j between (1) and (2):

$$-2\mu\lambda_{ij}T_{ij} + [g_i(M) + \mu]\,[1 + \lambda_{ij}T_{ij}] - [g_j(M_j) + \mu]\,[1 - \lambda_{ij}T_{ij}] = 0$$

and hence:

$$g_i(M_i)\,[1 + \lambda_{ij}T_{ij}] = g_j(M_j)\,[1 + \lambda_{ji}T_{ji}]$$

This result is equivalent to that proved above, namely that the incremental cost of increasing the load at node N_i is equal whether the extra generation is located at N_i or not.

8

HOURLY OPTIMIZATION

Given a solution along the lines discussed above, the marginal fuel cost and hence the marginal generating cost is given for each set in the system. Even if such a solution has not been obtained and the pattern of fuel flows has been determined in some other way, each set will have a given marginal generating cost if the fuel it uses is valued at a given marginal replacement cost. Henceforward it will be assumed that marginal costs have been determined in some way or other for each set and are known.

If transmission constraints and transmission losses are neglected, optimal operation is very simple so far as the purely economic aspect is concerned. At any given hour the system has a certain number of generating plants. Some of these will not be available for generation because they are undergoing repair or overhaul. Those which are available can be arranged in merit order, i.e. according to their marginal generating costs. The load required during the hour in question can be most cheaply met by running the highest merit (lowest marginal cost) plant, then the next and so on until total generation equals the given load.*

Even in the absence of transmission constraints (which are disregarded here, along with transmission losses) the plant available during any hour may not always be run in full merit order. There are several reasons for this, the most common being:

(*a*) At certain times of day, demand rises faster than the rate at which a conventional thermal plant can be brought up to full load;

(*b*) there is a minimum time of several hours for which it is economic to shut down conventional thermal plant; if demand is expected to rise again within a shorter period it is cheaper to run at minimum load and correspondingly reduce the load on other plants with slightly lower marginal cost than to shut down and then use heat in bringing the plant up to speed;†

---

* Where, as in the US, most sets are built with multiple valves and have a rising marginal cost curve, sets are not block loaded but are run at partial output for long periods. Thus a reduction in load is met by partially unloading a number of sets rather than by shutting down the lowest merit set or sets.

† If the rough approximation of covering these considerations by including an allowance for them in the 'performance factor' referred to earlier is to be avoided, economic despatching becomes considerably more complicated than simple use of a merit order. See L. L. Garver: 'Power Generation Scheduling by Integer Programming—Development of Theory', *IEEE Transactions, Power Apparatus and Systems*, February 1963.

(*c*) the unused capacity of the plant with the highest marginal cost among those running may be less than is desired to meet the risk of a sudden increase in demand (such as occurs at the end of a popular TV programme) and the risk of a sudden outage of generating plant. It will then be necessary to have several plants running with spare capacity to provide the required reserve.

These complications are all important in day to day system operation, but play a much smaller role in the context of broad system planning and in the costing of load increments which form the centre of attention in this book.

Once transmission losses and constraints are brought in, the problem becomes much more complicated. The general condition for the optimum stated above is clearly valid; the difficulty resides in the conflicting requirements of accuracy on the one hand and practicability for hour by hour calculation and application on the other. Practice ranges from the use of a rough approximate penalty factor for each station in the system to on-line computer-controlled despatching. There is a vast technical literature on the subject* which I shall not attempt to survey. It suffices to have made clear what it is that economic despatching (sometimes called economic system operation) is aiming at. It is important to note, however, that cost minimization is not the only consideration: security of supply is also relevant. Thus the system must be operated so that it can cope with the sudden unforeseen outage of a generating set or transmission link. Like the calculation of losses, this is a highly technical matter and can be dealt with at various levels of sophistication.

A final remark which is worth making is that decision-making does not have to be centralized. Two separately owned interconnected systems can determine the desired power flow between themselves by each quoting marginal cost schedules as forecast for a few hours ahead. The direction of power flow can then be planned to run from the system or region quoting the lowest marginal cost and (subject to any transmission constraints) to be of that magnitude which will make:

$$m^i = \frac{m^x}{1-\alpha}$$

* A formulation in linear programming terms and an extensive bibliography is given in S. Fanshel and E. S. Lynes: 'Economic Power Generation using Linear Programming', *IEEE Transactions, Power Apparatus and Systems*, April 1964.

While this use of a price mechanism* thus enables decentralized decision-making to co-ordinate the operation of interconnected systems, it does so only as regards hour to hour despatching. The planning of interconnection and the securing of economies in capital cost from it demands a unified approach.† Even the hour to hour co-ordination may not achieve the optimum which centralized despatching would provide, partly because contractual power exchanges may introduce rigidity and partly because load flow and security considerations introduce a non-cost dimension into optimization.

* A description by an economist is to be found in F. M. Westfield: 'Marginal Analysis, Multi-Plant Firms and Business Practice: An Example', *Quarterly Journal of Economics,* vol. 69, 1955.

† Thus the Federal Power Commission's *National Power Survey* states that 'The greatest total savings to all power users can be achieved only through planning which looks far beyond the requirements of a particular system or locality', p. 173, Washington, 1964.

2

THE OPTIMAL PLANT MIX
IN THE PLANT PROGRAMME

COST MINIMIZATION

Having discussed some aspects of the operation of a given generation and transmission system to meet a given load, we now turn to the problem of planning the expansion of the system to meet a given growth in load. Thus the analysis once again concerns cost minimization subject to various constraints relating, for example, to the security of supply, site availability, amenity and so on. Little will be said about these constraints here, since they are not really matters which can be discussed in generalized terms, except for security of supply. This brings in problems of uncertainty concerning future load growth and plant availability, and these problems are postponed for examination to a later chapter.

Minimizing the costs of meeting a given growth in load obviously involves thousands of design and construction problems. What is considered here is only broad system planning, which involves deciding what types of plant to install and where they should be installed in a given future period. If the planning, authorization and construction of new nuclear and conventional plant takes up to five years and if planning is an annual exercise, this means that what is at issue is the plant programme for plant to be commissioned during the year beginning five years ahead. The plant programmes for the intervening years are thus taken to be already decided and in train so far as nuclear and new conventional plant are concerned. Decisions to build gas turbine plant or to scrap old conventional plant can, however, be reconsidered during this period in the light of changing circumstances.

The analysis is strategic rather than tactical; it relates to the broad outline of the programme and leaves out the details of design and siting. These details, which are more highly technical, need to be

12

resolved once the main features of the programme have been determined. They are not discussed here.

In principle, there is of course a feedback between the forecast load growth which it is planned to meet and the plant programme. Costs depend on the programme and help to determine tariffs, which in turn affect the growth of the load. To bring this interdependence into a formal analysis of optimizing the plant programme would not be helpful, however, since it is impossible to take account of it in practice, simultaneously determining future tariff levels and the plant programme. It is just not possible either to predict the demand relationship between load, tariffs and other fuel prices several years ahead or to predict what price relativities will then be. Thus the feedback can only be taken account of implicitly and iteratively: each year's plant programme aims to meet a load which is forecast in the light, among many other things, of anticipated trends in electricity tariffs relative to the prices of other fuels; these anticipated trends in turn reflect the expected development of costs, depending, among other things, upon investment decisions already taken. Each time a new forecast is made for, say, six years ahead it may be different from the forecast made the previous year for seven years ahead.

This justifies the statement of the problem: choose the outline plant programme which will minimize the cost (subject to various constraints) of meeting a given forecast load some years ahead. It remains only to explain that by cost is here meant the present worth of all future net expenditures on the construction, maintenance and operation of the generating and transmission system.

FUEL SAVINGS

An essential concept which now has to be explained is the present worth of the fuel savings to be had from an extra kW of capacity of new plant with a given marginal generating cost.

For any given hour, the merit order can be represented as a rising marginal cost curve. In a large system, the possibility that it is stepped loses importance, so we can treat it as continuous in all cases. Let the marginal cost of the load met in that hour be m per kWh.

Now suppose that a new plant had been available during that hour with a marginal generating cost of r per kWh. If $r < m$, so that the new plant was higher in the merit order than the marginal plant, the given load could have been met more cheaply by running the new plant at maximum output and correspondingly reducing generation at the margin while maintaining output at all intermediate plants. The resulting reduction in system fuel costs for the hour in question would then have been $m-r$ per kW of the new plant's capacity.

In principle, such a calculation can be done for each of the 8,760 hours of each of the years of the new plant's anticipated life. If this anticipated life is thirty years the required data are thus the loads to be met, the marginal cost curves and whether or not the new plant would be available in each of 262,800 hours. The marginal cost curve in any hour depends on the capacity then existing (much of which has yet to be built) its availabilities, its heat rates and its marginal costs of fuel. The present worth of the saving for all these hours per kW of new plant can be represented as:

$$PW[(m-r)a]$$

where PW and the square brackets means 'present worth of the time-stream of' and where a is unity when the new plant is operating and zero when it is not—i.e. when $r>m$.

In practice, the present worth of the fuel savings per kW of a new plant can be estimated by computer simulation, hour by hour, of, say, eight representative days per year with and without the new plant given assumptions about the future shape of the system, fuel costs and the development of the load. The result can then be grossed up to yield a figure for the whole year. A repetition of the calculation for, say, every fifth year with a new set of assumptions each time, and interpolation for intervening years then provides a thirty-year time series of fuel savings which can then be discounted to obtain its present worth.

The necessary assumptions for more distant years are necessarily highly arbitrary. The operation of the discount factor reduces their importance, however. Furthermore in an expanding system it is likely (at least for conventional plant) that the bulk of the fuel savings will be concentrated into the first four to eight years of its life. Except in the case of the first sets of a new generation of plant, this is the part of the thirty years for which the background assumptions are least arbitrary.

The reason for the concentration of fuel savings in the early years is that new conventional plant will become less and less 'new' as time passes. It will lose its position in the merit order if plant installed subsequently has a lower marginal cost. This will occur if technical progress raises thermal efficiency or if different and cheaper fuels are introduced, e.g. nuclear generation. The resulting shift of the 'new' conventional plant lower down the merit order will make its annual fuel savings fall through time because:

(i) there will be more and more hours in each successive year when higher merit plant can meet the load, i.e. $r > m$

(ii) in the remaining hours, when $r < m$, m may fall through time as old, high-cost plant which was generating in those hours is superseded by less old plant with somewhat lower running costs.

The first effect will thus be to make $a = 0$ more and more frequently, so that the 'new plant' runs for fewer hours per year as time passes. The second effect will be that $(m - r)$ gradually falls.

The calculation of the present worth of the fuel savings is thus fairly insensitive to the assumptions made concerning the remoter years of the anticipated life of the new plant. Nonetheless, assumptions have to be made covering the whole period in order to derive an answer. We shall return to them later. Meanwhile, the point has been made that, given such assumptions, the present worth of the fuel savings from an extra kW capacity of new generating plant, PW $[(m - r)a]$, is a function of r, the marginal running cost of that new plant and that this function can be estimated by making the calculation for several values of r and interpolating. Note that r may change through time if fuel prices change. If such a change is expected, the calculations become more complicated without, however, introducing any new points of principle.

OTHER GENERATING COSTS

We have so far only discussed those costs which vary with the number of kWh generated, given the array of plant, i.e. fuel purchase, transport and handling and some repair and maintenance costs. The rest of system costs may be divided into three categories:

(i) capital cost of plant, including interest on construction, denoted C

(ii) the annual maintenance and manning costs of keeping plant in operation—i.e. what would be saved by scrapping it;* we call them Fixed Other Works Costs, denoted F

---

* In practice these will be greater for three-shift operation than for two-shift which, in turn, will be greater than for single-shift. To this extent these costs are related to the number of kWh generated, a complication which is ignored in what follows as is the complication that they rise with age.

(iii) system overheads: administration, research, etc. Since these are unaffected by the changes in plant mix considered here, they are not relevant to the analysis.

The net *change* in the present worth of all system costs arising from the addition or subtraction from a year's plant programme of 1 kW of generating capacity of any given type, given the load to be met in all future years, can now be seen to be:

$$C - PW[(m-r)a] + PW[F]$$

The present worth of fuel savings, it will be noted, enters with a negative sign since they act as a partial offset to the capital and fixed other works costs, C and F.*

The net change in the present worth of all system costs as defined above can be calculated for each of the various types of plant which are candidates for inclusion in the plant programme for the year concerned. It will be assumed here that there are four such types:

(i) New gas turbines, which have a low capital cost per kW, high marginal cost per kWh† and low fixed other works costs per kW

(ii) existing plant inherited from the past which can be scrapped, thus entering the plant programme as a negative item. At the margin such plant has practically zero opportunity cost per kW (i.e. scrap value and the value of the site), marginal cost per kWh ranging up to very high levels and fixed other works costs also ranging up to very high levels

(iii) new conventional thermal plant, which has a high capital cost per kW, low marginal cost per kWh and medium fixed other works costs per kW

(iv) new nuclear plant, which has a very high capital cost per kW, an extremely low marginal cost per kWh and high fixed other works costs per kW

THE MARGINAL CONDITIONS FOR OPTIMALITY

We can study the characteristics of the optimal plant mix of the programme for any year by examining the marginal conditions for

* Another minus item, omitted here for the sake of simplicity, is that new plant with a very low start-up time, such as pumped storage and some gas turbine plant, enables a reduction to be made in the amount of hot standby reserve. This saves heat costs.

† Since the maintenance of gas turbines is required according to the number of hours run, maintenance costs contribute to their marginal running costs.

optimality. These are that at no time shall a small substitution between any pair of the four types of plant reduce the present worth of all system costs. Since a decision to scrap old existing plant can be taken for a year at a time and since different types of new plant have different anticipated lives, it is convenient to express the conditions in terms of annual and annuitized costs. Thus using the notation $A\{\ \}$ to denote 'the annuitized value of ...' we can write the levelized annual change in system costs from a 1 kW change in capacity, given future loads, as:

$$A\{C-PW[(m-r)a]\}+F$$

and the marginal conditions are that this shall be the same for all three types of new plant and for the old existing plant which is just on the margin of being scrapped. If it is not the same, a change in the composition of the plant programme would reduce the present worth of future system costs.

Actually, it is not quite as simple as this because substitution at the margin would not be kW for kW owing to differences between the availability at peak of different types of plant.* Thus if the mean availability at peak of new conventional thermal plant is 90% while it is 95% in the case of gas turbines, a 1 kW reduction of new conventional plant in the plant programme must be replaced by a 90/95 kW increase in gas turbine plant. We shall therefore measure costs per kW of peak-available capacity so that, for example, the figures used for gas turbines are 100/95 times the cost per kW of installed capacity.

Subject to this understanding, we have the marginal conditions that:

$$A\{C-PW[(m-r)a]\}+F$$

be equal for all four types of plant. Since the present worth of fuel savings, $PW[(m-r)a]$ is a function of r, the marginal running cost of the type of plant in question, satisfaction of the marginal conditions can be stated as the requirement that the points relating to each of the four types all lie on the same isocost line in the following diagram. The line shows those combinations of r and $A\{C\}+F$ which involve the same $A\{C+PW[(m-r)a]\}+F$. Existing old plant which should be scrapped will be on points outside this line, while the

---

* In this chapter only mean availability is considered, dispersion round the mean being postponed for consideration in the chapter on risk.

larger amount of less old existing plant which should be retained will lie inside. Thus point z represents a nearly-new conventional plant whose running cost is only fractionally higher than that of new plant of the same type but whose construction cost is now a bygone and thus irrelevant to choice, its opportunity capital cost, *C*, consisting only of scrap and site value.

The above argument and diagram assume that the optimal plant programme will include some of all three types of new plant and some scrapping of new plant, in order to simplify the exposition. Sometimes, however, this will not be the case. Thus if, for example, the point for new nuclear plant lay to the right of the line even with

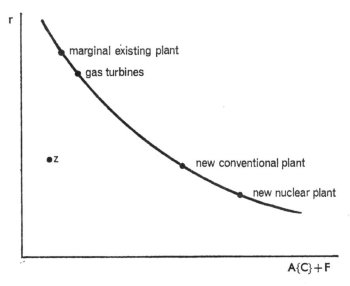

nuclear power occupying the smallest feasible share of the plant programme for the year, the optimal programme would obviously contain no nuclear component. Again, a rapidly expanding system may have no old plant which is ready for scrapping. A more formal statement of the marginal conditions would introduce inequalities to allow for such possibilities.

Strictly speaking, the marginal conditions as expressed above hold only for comparison between different types of plant whose anticipated lives are equal. This is because it is generally expected that technical progress in the design and manufacture of generating plant will continue in the future. Thus suppose that a comparison is to be made over a thirty-year period between plant A with an anticipated life of twenty years and plant B with an anticipated life of thirty years. Technical progress may be expected to make both the capital and running cost of the plant installed to replace A twenty years later lower than it was to start with. If these starting values are used to calculate the (twenty-year) annuity of its capital cost and the (thirty-year) annuitized present worth of its fuel savings, the cost of using and replacing A will be exaggerated relative to the cost of using B.

In principle, the right answer is to be found by extending the calculation to cover sixty years, the lowest common multiple of the lives, i.e. by comparing three successive type A plants with two successive type B plants. Uncertainty and the attenuation of remote considerations which is produced by the operation of the discount factor both suggest some cruder procedure, however. We shall consider one possibility at the end of the next chapter.

SECOND-ORDER CONDITIONS

We now have to show that equality of $A\{C+PW[(m-r)a]\}+F$ between those types of new plant which are included in the plant programme (and marginal existing plant) guarantees an overall cost minimum. Thus it has to be demonstrated that any change away from the optimum plant mix will raise the present worth of all system costs more and more the larger is that change. It will suffice to do this intuitively rather than rigorously by examining just two types of change.

Consider, first, scrapping less existing old plant and building fewer new gas turbines than is optimal. Successively increasing retentions of old plant would clearly raise costs progressively; such plant provides negligible fuel savings and so the various units can be ranked in increasing order of $A\{c\}+F$. Successively larger reductions in the gas turbines programme, on the other hand, will reduce capital and fixed other works proportionately, since these costs are uniform for all new gas turbines (apart from differences in local site conditions) and may raise the relatively small fuel savings from new gas turbines, since the more old plant is retained instead, the higher will be m at times of peak when the old plant has to be run.

Consider, secondly, building more gas turbines and less new nuclear capacity. The following diagram shows the system marginal cost curve as it is expected to be in a typical hour in the programme year. The continuous curve corresponds to the optimal plant mix and the dashed curve to a mix with more gas turbine capacity (running cost or_g) and correspondingly less nuclear capacity (running cost or_n). At an hour when the load is *od*, the fuel savings from a marginal kW of available nuclear capacity will be:

$dm - or_n$ with the optimal mix

$dm^1 - or_n$ with too much gas turbine capacity and too little nuclear

It is clear that the excess of dm^1 over dm will be greater the farther apart the two curves, i.e. the greater is the deviation from the optimal

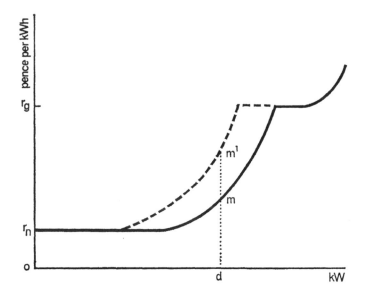

plant mix. This means that as more and more gas turbine capacity is substituted for nuclear capacity, the fuel savings sacrificed by reducing nuclear capacity will grow. The excess of the capital and fixed other works costs of nuclear plant over those of gas turbines will remain unchanged, however. Thus the increase in total net costs arising from the substitution will increase continuously as the substitution moves the mix further and further from the optimum.

NON-OPTIMAL PLANT MIX

There are several reasons why it may not be possible to secure an optimal plant mix in the programme for a year. One is that the calculation of such an optimum may be an impossibly large task once all the practical complications are brought into account. While standard costs are required for such calculations, each particular site presents its own problems and causes cost differences from case to case. Furthermore, the general calculation ignores geography and hence transmission costs which are, of course, extremely important.

Secondly, there is an initial difficulty with new types of plant where operating experience is lacking and cost estimates are consequently hazardous. Caution urges that such plant be used less than appears optimal.

Thirdly, at times when a new technology has come to fruition it may be impossible to take full advantage of it in one year. Thus the total which it would be desirable to install may exceed one year's load growth or may require more manufacturing capacity than is available.

Fourthly, future load forecasts are revised as time passes and new information is acquired. Hence provisional plant programmes have to be revised. Since it takes less time to close down existing thermal plant than to plan and construct gas turbine plant and this in turn takes less time than the planning and construction of new conventional plant, upward or downward revisions of plant programmes may make the composition of the final programme different from what it would have been had there been no change in the load forecast. Thus, to take the extreme case, a downward revision of the estimate of capacity required within the next two years, can only be met by accelerating the retirement of old plant or possibly by slowing down (rather than reducing) the construction of new plant. The latter, however, is frequently regarded as undesirable on the grounds that the postponement of the fuel savings which will start to provide when it is commissioned will outweigh the saving in interest.

This last point brings in the problem of uncertainty in forecasting, which clearly has an impact upon the problem of deciding how much capacity to plan for. This problem is taken up later, in another chapter, however, and for the present it is sufficient to note that optimality cannot always be achieved.

TRANSMISSION COSTS

It is now time to bring in transmission costs, i.e. to give the analysis a spatial dimension. This involves three complications: first, there is the capital cost of transmission capacity; second, there are transmission losses and, third, there are the security constraints resulting from the requirement that various specified possible outages of generating or transmission capacity shall not lead to interruptions of supply. The following discussion ignores the third of these complications and concentrates on the point that the expenditure of extra capital on transmission makes possible increased transmission of energy from low cost to high cost nodes with a consequent saving in generation costs but with some extra transmission losses. This, as noted below, is not meant to imply that such saving constitutes the main justification of transmission. For simplicity it will be assumed, as in the preceding chapter, that there are only two nodes, x a low-cost node and i a high-cost node.

Given the loads and generating capacity of each type at each node and given the transmission capacity linking them, optimal operation of the system requires that the incremental cost at all hours of providing an extra kWh in i by generating it there can only exceed the incremental cost of providing it by extra generation in x when the transmission line is loaded to capacity. Thus we can write:

$$m^i = \frac{m^x}{1-\alpha} + \varepsilon$$

where $\alpha=$ incremental transmission losses which are a function of the type and length of the transmission line and of the load, so that α varies from hour to hour. The variable ε is zero except at times when the transmission line is loaded to capacity. When it is, i.e. when the transmission constraint is operative, ε takes whatever positive value is required to preserve the equality. ε can thus be interpreted as a shadow variable equal to the excess over (i) what m^i would have to be to make the optimal power flow equal to the capacity of the line of (ii) its actual value.

It is now obviously necessary to introduce a marginal condition for optimality relating to the transmission capacity requisite to minimize total system costs. The condition here, for a given load and generating capacity at each node (and neglecting any fixed other works costs of transmission), is that the capital cost of a unit increment of transmission capacity should just equal the present worth of savings in running costs which could be obtained with that unit increment in transmission capacity.* This can be stated as follows:

$$C_t = PW[m^i dQ^i - m^x dQ^x]$$

where Ct is the unit capital cost of extra transmission capacity, measured at the transmitting end, and dQ is the change in output generated. Since the total load in x and i is given, dQx must exceed dQi by the increase in transmission losses dL:

$$dQ^x = dQ^i + dL$$

As an approximation, dL can be regarded as the algebraic sum of (i) incremental losses with respect to load, given transmission capacity, i.e. α, multiplied by the increase in load and (ii) the reduction in losses with respect to transmission capacity, given load, denoted β:

$$dL = \alpha\, dQ^x - \beta$$

Substituting the last two expressions into the transmission capacity marginal optimality condition gives:

$$C_t = PW[dQ^x(m^i - \alpha m^i - m^x) + m^i\beta]$$

but since merit-order operation keeps $m^x = (m^i - \varepsilon)(1-\alpha)$, this reduces to:

$$C_t = PW[m^i\beta + dQ^x \varepsilon(1-\alpha)]$$

This means that the capital cost of a unit increase in transmission capacity must equal the sum of the present worths of the two kinds of saving in running costs obtained from it, namely:

(*a*) The reduction in the transmission losses consequent upon the increase in capacity, given load, saving $m^i\beta$ in each hour;
(*b*) The substitution of 1 extra kWh generated in x for $(1-\alpha)$ generated in i at times when the line is loaded at capacity, i.e. when $\varepsilon > 0$. Since capacity is increased by 1 kWh, dQ^x at such times equals 1 kWh. The saving in i is $m^i(1-\alpha)$ and the extra cost in x is m^x, so the net saving is $m^i(1-\alpha) - m^x$. Substituting

$$m^i = \frac{m^x}{1-\alpha} + \varepsilon$$

gives $\varepsilon(1-\alpha)$ or, since $dQ^x = 1$ when $\varepsilon > 0$, $dQ^x\varepsilon(1-\alpha)$.

* If security considerations were not being omitted from this simple analysis, it would be necessary to add to the saving in running costs a term to account for the saving in reserve requirements of generating plant resulting from the increase in transmission capacity.

It can be seen that as transmission capacity increases, given the load, both β and ε will fall (while α will not change at full capacity load). Hence given all the other variables in the system, there will be a unique optimum transmission capacity unless C_t varies discontinuously.*

The argument has proceeded as if i never exports to x, but this is not required by the algebra, where which node is denoted x and which i can be allowed to change according to which node is the exporting one. It is, after all, quite realistic to suppose that the direction of power flow between two nodes will not always be the same. For example any node with a lot of its generating plant out of service for maintenance will have to import at such times even if it is generally an exporting node on account of low fossil fuel costs. Again, a node with a large nuclear capacity may export power off-peak, thus bearing a large share of the system's base load, but import power on-peak from nodes which have a lot of low-merit plant.

More important still, different nodes may have their peaks at different times. Such diversity, i.e. the fact that peak system load is less than the sum of the separate peaks at all the nodes of the system, is in fact a major source of the economies provided by interconnection. Total generating capacity is less than it would have to be in the absence of interconnection both for this reason and because of the economy in reserve capacity provided by pooling. Thus the above discussion of the marginal conditions is not meant to imply that power transfers designed to economize on running costs constitute the main justification of interconnection. The point is merely that once interconnection exists, for whatever reason, the marginal conditions for the optimum include the above relationships between running costs and the capital cost of transmission.

PLANT LOCATION

The marginal optimality conditions first stated related to the plant mix at a single node, given the future loads to be met. We have now assumed given generating capacities and given future loads to be met at each of two linked nodes and derived an additional optimality condition relating to the capacity of the transmission link between them.

* In practice, it may well do so, since voltage, line size and the number of lines are not continuously variable.

These marginal conditions all take the location of capacity as one of the givens. To deal with this an additional set of marginal conditions is required, relating to the allocation of capacity between nodes given the transmission capacity linking them. This is the obverse of the problem examined in the preceding section.

If the plant programme includes all three types of new plant and some scrapping of old existing plant at each of a pair of nodes, there are sixteen possible kinds of inter-area substitution. Each of these involves substituting 1 extra kW of available capacity of a given type in node 1 at an extra cost of:

$$A\{C_1 - PW[(m_1 - r_1)a_1]\} + F_1$$

for 1 less kW of available capacity of a given type at node 2 at a saving of:

$$A\{C_2 - PW[(m_2 - r_2)a_2]\} + F_2$$

assuming that transmission capacity is not used to capacity for import into node 2 at the time of peak load there so that 1 kW extra can be imported or 1 kW less exported.

Such a marginal transfer of capacity between the nodes will not sensibly affect marginal generating costs in either, so the pattern of generation in each and of power flow between them is treated as unchanged. Thus the net effect on costs is confined to the algebraic sum of the two expressions above and this must be zero if the existing location is optimal:

$$A\{C_1 - PW[(m_1 - r_1)a_1]\} + F_1 - A\{C_2 - PW[(m_2 - r_2)a_2]\} - F_2 = 0$$

for each of the sixteen comparisons.

The pairwise comparison of two nodes in this fashion is only possible if there are no other indirect links between them. If there are such links, i.e. if the network contains loops, the consequences of a marginal transfer of capacity are extremely complex.

SUMMARY OF THE CHAPTER

A number of implications of cost minimization, namely that none of the various conceivable changes in the plant mix should raise the present worth of total system costs, have been set out as marginal conditions for optimality. It has been shown that (except for the transmission condition where discontinuities may complicate matters)

25

the second-order conditions are likely to be fulfilled if the marginal ones are.

The marginal conditions are of interest in two contexts. First, they could serve to test the optimality of any specific plant programme which is put forward. They could not be used to derive an optimal programme directly but would indicate the direction of change required to improve a non-optimal programme. They could thus serve as a means of checking economic efficiency.

The second way in which the marginal conditions are potentially useful will, unlike the first, be developed further in later chapters. This is in the analysis of the structure of marginal costs, a matter of central importance for optimal pricing policy.

3

THE BACKGROUND PLAN

The marginal conditions for an optimal plant mix have been presented both because they are of use in studying the cost structure of electricity supply and because they can serve to check upon the optimality of a proposed plant programme for a particular year. In either case they involve the calculation of the future time stream of fuel savings to be had from new plant. Now it is clear from the description of how this time stream can be calculated that the calculation requires knowledge—or assumptions—about the future composition of the system. This is because the shape of the marginal cost curve in future years will reflect the type of plant which is available in those future years. So the fuel savings relevant to the choice of plant for, say, 1975 can only be calculated if the plant available in 1976 and subsequent years is known. This means that the plant programme for 1975 can only be optimized if the plant programmes for subsequent years are already known.

The way out of the apparent logical impasse is to have a two-stage optimizing procedure. First the broad outlines of the pattern of system development over the twenty-five to thirty years beginning in 1975 must be determined. Second, in the light of the background plan so obtained, more detailed consideration is given to the plant programme for 1975 and firm decisions are made. The background plan thus does not itself constitute a set of decisions but is an essential pre-requisite for the second stage which does lead up to decision-making.

It is a natural thought that the two stages might be linked in an iterative analysis. Whether this is possible or not will not be discussed here. But it is clear that a background plan can be anything between the quantitative expression of expert guesstimates on the one hand

27

and the product of elaborate and computerized calculations on the other. The uncertainty concerning future technical developments, relative fuel costs and the development of the load can be dealt with by hunch or by refined parametric analysis of the optimum.

However the problem is tackled in practice, it seems worth while presenting a very simple model here. This only has didactic value, but is necessary in order to show how the main relevant factors interact. The analysis of the last chapter dealt with the marginal characteristics of the optimum but gave no idea about its general or total characteristics, such as the determinants of the share of nuclear power in the total.

SIMPLIFYING ASSUMPTIONS

As economists have found in other contexts, the assumptions of uniform growth and the absence of history powerfully simplify a dynamic analysis—by transforming it into a static one. So let it be assumed that:

(*a*) there are, and always will be, only three types of plant: gas turbine, conventional thermal and nuclear

(*b*) relative prices and techniques are given and unchanging

(*c*) load is growing uniformly year in and year out with no change in its hourly pattern

(*d*) the mean availability for each type of plant over the whole year is equal to its peak availability.

Under these conditions, the optimal plant mix will be the same in each and every year and the marginal generating cost curve will be as shown in the accompanying diagram. Each year the curve will stretch sideways by the annual rate of growth of the load. Hence each of the three types of plant will be used in exactly the same way in each successive year and optimization can be looked at in terms of any one year.

ANALYSIS FOR A SINGLE NODE

We start by ignoring transmission. Hence the only problem is the proportion of available capacity constituted by each of the three types of plant.

Using g, c and n to denote gas turbines, conventional and nuclear plant respectively, the marginal conditions are

$$A\{C_g\}+F_g$$
$$= A\{C_c - P\overline{W}[(m - r_c)a_c]\} + F_c$$
$$= A\{C_n - P\overline{W}[(m - r_n)a_n]\} + F_n$$

Gas turbines produce no fuel savings since there is no plant in the system with higher running costs.

Now consider the fuel savings of conventional plant. m here is the marginal running cost of gas turbines, i.e. $m = r_g$. If the number of hours per year when demand exceeds the available capacity of

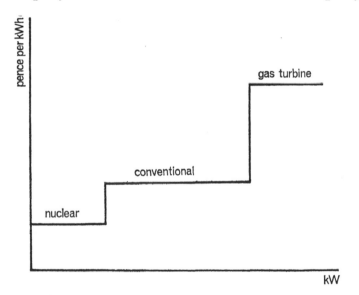

conventional and thermal plant is H_g, gas turbines will be run for H_g hours a year. So the annual fuel savings from an extra available kW of capacity of conventional thermal plant will be $(r_g - r_c)H_g$. The first equality above thus becomes:

$$A\{C_g\} + F_g = A\{C_c\} - (r_g - r_c)H_g + F_c$$

which gives:

$$H_g = \frac{A\{C_c\} + F_c - A\{C_g\} - F_g}{r_g - r_c}$$

In a similar manner, the m in the fuel savings term for nuclear plant is $r_g - r_n$ for H_g hours in the year and $r_c - r_n$ for H_c hours in the year, H_c being the number of hours in the year when demand exceeds the available capacity of available nuclear plant but is not so high as to require generation by gas turbines. For H_c hours in the year, in other words, system marginal cost (m) is r_c. Thus the second equality becomes:

$$A\{C_c\}-(r_g-r_c)H_g+F_c = A\{C_n\}-(r_g-r_n)H_g-(r_c-r_n)H_c+F_n$$

which gives:

$$H_g+H_c = \frac{A\{C_n\}+F_n-A\{C_c\}-F_c}{r_c-r_n}$$

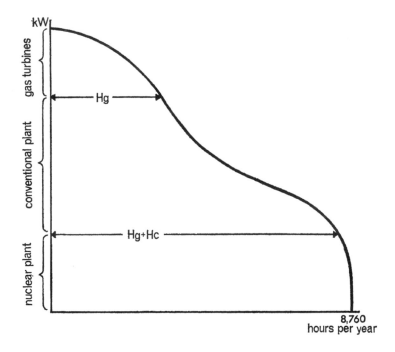

as the maximum number of hours in the year for which conventional thermal plant is run, the minimum number being H_g. Nuclear plant, finally, runs a maximum of 8,760 hours per year and a minimum of H_g+H_c.

The relationship between H_g and H_c on the one hand and the plant mix on the other can now be shown diagrammatically with the aid of a cumulative load-duration curve. This shows for any given load, measured vertically in the accompanying diagram, the number of hours in the year, measured horizontally, when load equals or

30

exceeds that given level. Insertion into the curve of H_g and H_g and H_c gives the optimal composition of available plant capacity as shown on the vertical axis. It also divides the area under the load-duration curve into parts each showing the total number of kWh generated per year by the type of plant in question. It is immediately apparent that kWh generated per kW of available capacity is highest in the case of nuclear plant and lowest in the case of gas turbines.

The proportion of gas turbines in the optimal mix clearly depends on the variables in the formula for H_g and on the shape of the load duration curve. *Ceteris paribus* it will be raised by:

— increased capital or fixed other working costs of new conventional plant relative to those of gas turbines
— increased average availability of gas turbines relative to that of new conventional plant
— an increase in the rate of interest, since new conventional plant has a higher capital cost per available kW
— increased running costs of new conventional plant relative to gas turbine plant
— a downward shift in the portion of the load duration curve at a duration of H_g.

Similar factors affect the share of nuclear plant in the optimal mix. It is plausible that these purely qualitative conclusions would presumably still hold even when the simplifying assumptions were not met, which makes them of wider interest than the simplifying assumptions might suggest. Thus if there is some existing conventional plant, the optimal mix will trend asymptotically to the level it would have if there were no such plant.

TWO NODES

When we consider a system consisting of two linked nodes, the choice of the optimal plant mix within and between each node and of the optimal size of the transmission link becomes very complicated. In order to continue with the simplified case of uniform growth, no technological progress and so on, it is necessary to adopt two further assumptions.

The first is that the only item of costs which differs between the two nodes is the running cost of new conventional plant. This, it may be supposed, is due to regional differences in coal or oil prices.

Since we are not descending to the finer details of site selection and thus regard the nodes as being fairly large areas, the assumption that all other costs are the same in both nodes is realistic. Apart from local site conditions, the cost of constructing and maintaining a plant in one area is much the same as the cost of doing it in any other area.

The second new simplifying assumption is that the loads at each of the nodes are identical. This is a presentationally convenient special case of the more general requirement that the ranking of the 8,760 hours in the year by kW of load is the same for each node. This is a 'requirement' because it gets rid of any economy which is to be had from linking the two nodes due to diversity between their loads. Important though this may be, we ignore it here in order to concentrate on one question only, namely the spatial aspect of the problem which involves regional differences in fuel costs on the one hand and transmission capacity and plant location on the other.

SCALE ECONOMIES

Before tackling this question, it is worth pausing to explain the nature of the other economies to be secured from interconnection. As mentioned in the last chapter, the diversity which we have just assumed away can enable interconnection to provide economies. This is readily apparent if we think of connecting a system with a winter peak to a system with a summer peak. The sum of their peak demands exceeds the peak of the sum of their demands, so that interconnection reduces capacity requirements.

A second economy in capacity requirements stems from the reduced percentage reserve margin of capacity over system maximum demand which is required when two separate systems are interconnected. As with diversity, this is a phenomenon of scale rather than of space *per se*, and has here been pushed outside the picture by the assumption that, for each type of plant, availability is known and constant throughout the year. In practice, the pooling of the risks of unplanned plant outages and the staggering of outages for planned maintenance are both important. The risk element is discussed in a later chapter.

Finally, there is an economy of scale which has been disregarded by the implicit assumption that new plant is divisible. In a large system, where one year's plant programme includes several large sets, this assumption does not falsify the picture. But a small system, with an annual required growth in available capacity less than the size of a single large set, has to choose between building ahead of demand in some years and installing sets which, being small, may have a higher capital cost per kW or higher running cost than the large set would offer. Thus there is scope for two neighbouring systems to secure the economies of large set size by having a transmission link and co-operating in the installation of large new sets.

The various scale economies just mentioned are inter-related. For example reserve plant margin is affected by the size of new sets currently being installed. The reason is that the probability of unplanned outage may not be the same for all set sizes, being lower for sets of a tried and tested type. (Note that the sudden loss of a large set requires a bigger margin of spinning reserve to cover it than does the sudden loss of a small set.)

We are leaving all these scale effects out of account, however, together with any organizational diseconomies. The economies just described provide a rationale for interconnection and pooling agreements to achieve some degree of co-ordination between separate systems. What we are concerned with here, is the optimal planning of a given system of given size and rate of growth. The purpose of the simplifying assumptions is to focus attention upon the spatial problem of regional differences in fuel costs.

TRANSMISSION

The problem now is to locate plant and choose the amount of transmission capacity so as to meet the given load at each node at minimum total cost for the system. This cost is the sum of:

— the annuitized capital cost and annual fixed other works cost of each of the three types of plant at both nodes
— the annuitized capital cost of the transmission link
— running costs

It will be realized that since the use of the transmission link involves incurring transmission losses, the total number of kWh generated in the system will have to exceed the aggregate of the two loads by the amount of these losses.

Despite the plethora of simplifying assumptions, the model is too complicated for algebraic or geometric solution. Such a solution is not an end in itself, however, as its purpose is not to serve as a practical tool but merely to indicate the factors which are relevant and how they affect the nature of the optimal position. It turns out that we can find out quite a lot about these factors without a formal solution by virtue of the fact that a solution for a finitely large level of transmission costs will lie somewhere between the solutions for the limiting cases of an infinitely large and an infinitely low level. The expression 'level of transmission costs' is here used to cover both the capital cost per unit of capacity and the transmission losses.

Consider first the case of infinitely high transmission costs. The optimal transmission capacity will then be zero and generation will

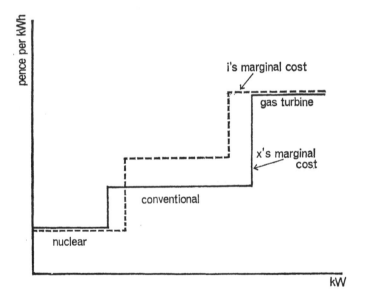

match the load at each node separately. The plant mix will differ between them, however, since one cost item, the marginal generating cost of conventional plant (r_c) is higher at one node, node i, than at the other, node x. Consequently:

H_g, the maximum number of hours per year for which gas turbines run, will be higher in i than in x

$H_g + H_c$, the minimum number of hours per year, for which nuclear plant runs, will be higher, in i than in x.

Since we have assumed that the size and shape of the cumulative load duration curve is the same at each node, this means that i

will have both more available gas turbine capacity and more available nuclear capacity than x so that x will have correspondingly more available conventional thermal capacity. Hence the marginal cost curves of the two nodes will appear as in the accompanying diagram. The fact that x has more available conventional thermal plant than i is shown by the greater length of the middle segment of its marginal cost curve.

Now consider the opposite case, that of zero transmission costs. i will now have no conventional thermal plant at all, since it is now cheapest to locate all such plant at x and have i import its requirements from x. This leaves the location of gas turbine and nuclear plant indeterminate, since the cost will be the same, on present assumptions, however these two types of plant are allocated between the two nodes. To resolve this, let us bring in the requirement that insofar as it can be achieved at zero cost, the available capacity installed at each node should equal peak demand at that node. The rationale of this is that it makes possible the continuity of supplies at both nodes in the event of a transmission outage.*

The marginal cost curves at the two nodes can now be shown as in the diagram overleaf. i has been allocated all the nuclear plant and practically all the gas turbine plant in order to have the same total available capacity as x. (Alternatively it could have been allocated all the gas turbines and practically all the nuclear plant. And if the optimal available capacities of conventional thermal plant had been a bit greater than shown i would have been allocated all of both the other two types and a smaller total than x.) The total amount of each type of plant in both nodes together is just twice what it was at node x alone in the previous diagram, since (with zero transmission costs) cost relativities for both nodes together are in this case the same as they were for node x alone in the first case.

Thus the net effect of shifting from a high to a low level of transmission costs is, first and most obviously, to have a greater transmission capacity; secondly, to have more conventional thermal plant in node x; thirdly, to have more gas turbine and/or nuclear capacity in i; and, finally, to have more conventional plant in total (and hence less of the other two types in total).

---

* In practice, of course, extra costs are incurred for the sake of increasing the security of supply. The analysis of this is left for a later chapter where uncertainty is introduced. Meanwhile the innocuous assumption that security is desirable if achieved at zero cost provides a convenient *deus ex machina*.

The effects upon the optimum of changes in the relationship between fossil fuel costs in x and i should now be apparent. A reduction in r_c in x will have the same effects as a reduction in the level of transmission costs. A rise in r_c in i will also increase the optimal amount of transmission capacity and involve more conventional thermal plant in node x but will obviously reduce rather than increase the share of conventional plant in the total capacity of both nodes taken together.

The final determinant of the optimum which deserves attention is, as in the single node case, the shape of the load duration curve(s).

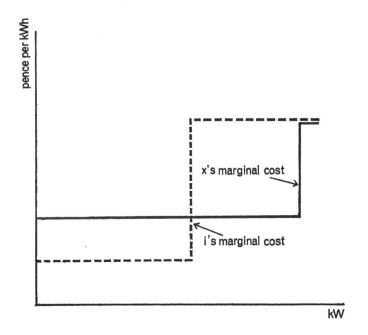

If the portion of this curve lying between H_g and $H_g + H_c$ shifts upward, the gain to be had from shifting some conventional thermal plant from i to x and from increasing transmission capacity is increased.

SIGNIFICANCE OF THE MODEL

This completes the analysis of the model. It has been designed merely to provide a qualitative understanding of the effect upon the optimal pattern of generating and transmission capacity of the most significant variables. These are the capital, running and fixed other works costs of different types of generating plant; the costs of transmission (which reflect distance); geographical differences in fuel prices and, finally, the shape of load curves. A change in any one of these items may well affect most features of the optimal pattern.

An interesting by-product of the discussion is the observation that transmission flows may not always be in the same direction, a point which was mentioned in the previous chapter. In both the two extreme cases pictured in the last two diagrams and hence in intermediate cases as well, there is a range of load where marginal cost is lower in i than in x. At such loads, any transmission will be from i to x, since the cheapest way to meet an increase in load in x will then be to increase nuclear generation in i. Thus the high conventional fuel-cost node will export power if there are times when total system load falls below the total available nuclear capacity of the whole system. This is to say that once available nuclear capacity exceeds base load (the level below which load never falls) the high fuel-cost node will export at off-peak times and then shift over to import as the load rises.

A PROGRAMMING APPROACH

Both the Central Electricity Generating Board and Electricité de France have devoted much effort to developing a programming approach to the formulation of what is here called a background plan. The latest French version is non-linear, partly on account of the rising cost of hydro-electric plants as successively less and less favourable sites are used and partly to allow the calculation of the risk and cost of interruptions of supply using a normal probability distribution. This complication is ignored here, however, and the following sketch is in linear terms. All that is attempted here is to indicate that the problem can be formulated in terms of linear programming, without going into the detailed problems and limitations of the application of the technique. In particular, we continue to ignore uncertainty.

The system is divided into areas (nodes) and the year into a small number of sub-periods, such as the peak hour, winter nights, etc. during each of which the load is constant. The load is forecast for each node for each sub-period for each year. This can be simplified by taking, say, only each fifth year.

Assuming that the only cost item to vary between nodes is fuel cost, the capital and fixed other works cost for each type of plant and its thermal efficiency is specified together with the cost of fuel at each node in each year considered. Technical progress can be allowed for by distinguishing sub-types of plant according to cost and/or thermal efficiency and imposing the constraint that the lower cost sub-types cannot be installed before a certain date. Existing plant is lumped into groups, each of which is treated as homogeneous and whose scrap value is treated as a credit on retirement. Plant outage for maintenance can be allowed for by multiplying the installed capacity of each type of plant by an appropriate co-efficient for each sub-period.

The total cost to be minimized is the sum of capital costs, fixed other works costs and running costs where each item is multiplied by a discount factor to give its present worth at the base date. The total running cost in each year considered is the sum for all nodes and types of plant of kWh generated times fuel cost per kWh.

The most obvious constraints, apart from the non-negativity ones, are, first, that for each sub-period for each year the load at each node shall not exceed generation at that node plus its net import and, second, that generation shall similarly not exceed available capacity.

The introduction of inter-node transmission brings in transmission cost, losses and constraints. Some approximation is required to express these linearly.

For present purposes this will suffice to give an idea of how a background plan could be determined with the aid of linear programming. A choice needs to be made between realism and manageability. However this is resolved, the result can never be a detailed capital programme. All that can be provided is a broad outline of optimal system development over the next decade or two. Indeed such is the uncertainty concerning, for example, the future development of nuclear costs, that a parametric analysis rather than a single solution is desirable. (This is what EDF do in their analysis.) The analysis is thus a tool for examining the implications of managerial judgements concerning these uncertain variables. It can never obviate the need for these judgements—except in the happy event that it shows results for the earlier years to be highly insensitive to the assumptions for later years.

The point is, then, that basic judgements (which can be revised later on) can be turned into an outline plan for the future development of the system. This plan is not itself a commitment to action. Its function is simply to provide a set of reasonable assumptions which make possible those calculations of fuel savings that are a necessary part of optimizing the next capital programme.

AN ALTERNATIVE APPROACH

The two-stage optimization discussed so far is very elaborate. It is perhaps worth pointing out the nature and the limitations of the simpler procedures adopted in many undertakings by considering one example. This conveniently serves to bring home the importance of technical progress as a key factor.

Let us start by recalling the nature of the problem. Given (i) the future growth of load; (ii) the shape of the load curve; (iii) the estimated future capital and fixed other works costs of generating plant and associated transmission which could be commissioned in future years; (iv) future fuel costs and, finally, (v) the plant margin, it is to decide now what plant to commission in some future year, year m. The required capacity of this plant follows from the load forecast for that year. The aim is to minimize the present worth of all system costs from now to infinity, discounting them at an appropriate discount rate.* This is an ideal rather than a practical criterion because:

(a) the uncertainty attached to load and cost estimates increases with their futurity
(b) the complexity and hence the cost of decision-making argues in favour of something simpler,

yet it has a value as a standard of reference by which practical approaches can be judged.

The particular practical approach now to be examined is to specify a number of technically feasible plant programmes covering generation and transmission for, say, the ten years commencing with year m. The plant programme chosen for year m is then the first year of that ten-year programme with the lowest costs. These are calculated as the present worth of the ten-year time stream of the sum of:

---

* Where the system is growing faster than the discount rate, the present worth will be infinite. But choice between two alternative schemes can still be made according to the present worth of the difference between their costs.

(a) annuitized capital cost of plant put in after $m-1$
(b) its annual fixed other works costs
(c) total annual system fuel costs.

This practical approach would be completely congruent with the (impractical) ideal if it were done for an infinite number of years instead of ten (though in this case there would be no point in first annuitizing and then re-capitalizing capital costs). Thus the weakness in principle of this approach is its disregard of costs subsequent to year $m+10$. Its strength is that it takes explicit account of any inter-dependence between the optimum choice for year m and that for subsequent years.

In order to examine further this weakness, let us go back to the ideal criterion and restate it as minimizing the sum of the present worths at the reference date of:

all system costs from m till $m+10$
$+$ all system costs from $m+10$ till ∞.

Now consider the present worth of what all system costs from $m+10$ to ∞ *would* be *if* no new plant were put in from m till $m+10$. Since these costs are independent of what is *actually* going to be done from m till $m+10$, they merely constitute a constant term. We can therefore subtract them without making any difference and restate the criterion as minimizing the sum of the present worths at the reference date of:

all system costs from m till $m+10$
$+$ all system costs from $m+10$ till ∞
$-$ what all system costs would be from $m+10$ till ∞ if no new plant were put in from m till $m+10$.

This restatement enables us to get a simple result for the special case where no future technical progress is anticipated, i.e. where the capital, the fixed other working and the running costs of each alternative type of new plant (coal, nuclear, etc.) are unchanged in all future years. For under these conditions the sum of the present worths of the capital cost components of the three items in the expression at the end of the last paragraph equals the present worth of the annuitized capital cost over the years from m till $m+10$ of the new plant put in during the period.

To demonstrate this, consider a simple case where only one item

40

of capital expenditure is required between m and $m+10$, namely £C in year $m+5$, and where the economic life of this one item is thirty years. It follows from this expository assumption and from the general assumption of zero technical progress that:

(a) 'all system costs from m till $m+10$' include only one capital item, C in the year $m+5$

(b) 'all system costs from $m+10$ till ∞' will include a replacement expenditure of C in each of the years $m+35, m+65 \ldots$

(c) 'what all system costs would be from $m+10$ till ∞ if no new plant were put in from m till $m+10$' will include the expenditure of C in each of the years $m+10, m+40$ and $m+70 \ldots$

Let us express these costs as equivalent thirty-year annuities. Then (a) plus (b) is equivalent to a perpetual annuity beginning in year $m+5$, while (c) is equivalent to a perpetual annuity beginning in year $m+10$. Hence $(a)+(b)-(c)$ (which is the equivalent of the capital expenditure component of the expression to be minimized given on the opposite page is an annuity of the same amount extending from the year $m+5$ till $m+10$.

This shows that the approach described would only be consistent with the ideal criterion if future technical progress were assumed to be zero. There would then be no difference between (i) all *other* system costs from $m+10$ till ∞ and (ii) what they would have been if no new plant were put in from m till year $m+10$, as the hypothetical plant that would then be put in for $m+10$ would be identical with that actually put in for the decade. However the approach does in fact recognize technical progress in the years examined— up to year $m+10$—so is not formally consistent with the ideal criterion.

The question therefore arises whether any modification of the approach would bring it nearer the ideal. So let us return to the minimand, namely the present worth of:

all system costs from m till $m+10$
+all system costs from $m+10$ till ∞
−what all system costs would be from $m+10$ till ∞ if no new plant were put in from m till $m+10$.

The algebraic sum of the present worths of the last two items is simply minus the present worth of the residual value at $m+10$ of the plant which will be put in from m till $m+10$. Thus the minimand is simply the present worth of costs incurred during the decade less the residual value of plant at the end of it. This residual value will be less than what that plant cost when it was put in not only because, as the approach recognizes, part of its useful life has expired, but also because technical progress will have reduced the cost of the alternative new plant which its possession saves. The difficulty with the constant annuity used in the approach is that it fails to recognize this second reason. The trouble is that a complete recognition of it would require an extension of the analysis to ∞.

It is worth considering, however, whether some rough recognition of the effect on its residual value of technical progress occurring after plant is installed during the ten-year period would not be better than none. Certainly it seems inconsistent to recognize it as affecting the cost of new plant installed during the course of the years up to $m+10$ but implicitly to deny it when comparing the ten-year period with conditions at its end. Such recognition is easy as regards capital cost and fixed other works costs. Thus:

(a) If capital cost per kW of a plant to be installed for the year $m+10$ is expected to be, say, £35 as against £38 in year $m+5$, and if life expectancy is thirty years, the residual value per kW in year $m+10$ can be taken as the present value then of the first twenty-five years of the thirty-year annuity corresponding to £35;

(b) if fixed other works costs per annum per kW are expected to be, say, £0·40 as against £0·43, the difference (£0·03) can be subtracted from the above annuity.

This is rather crude since, by virtue of the same kind of argument as that given above, it neglects technical progress between $m+35$ years and $m+40$ years hence, but the operation of the discount factor attenuates the significance of this factor.

A similar rough way of taking account of that technical progress which takes the form of reduced running costs is more difficult to find. This kind of technical progress affects residual value via its effect on fuel savings, and these cannot be calculated without assuming that knowledge of the state of the system in the years $m+10$ onward which (*ex hypothesi*) is lacking. Nonetheless, if a plant installed for $m+10$ would generate at a lower running cost per kWh than a plant installed for year $m+5$, merely following (a) and (b) above would overstate the residual value of the latter in year

$m+10$. Perhaps some estimate of this overstatement could be made, and applied to modify the residual value as calculated by (*a*) and (*b*) above, by examining the effect which the same running cost difference would have in the years *up to* $m+10$. These can be simulated.

FINAL REMARKS

This chapter simplified the problem enormously in order to derive some qualitative results which probably remain valid under the complex conditions which face planning engineers in reality. A background plan, whether formulated crudely or with the aid of a large programming model, should thus conform with the principles enunciated. However, the two-stage approach, iterative or not, between a background plan and the plant programme for a given year is not the only possible approach. A single-stage approach which may be particularly suitable for smaller systems has been described. The relevance of anticipated technical progress to planning decisions was demonstrated in terms of this single-stage approach, such progress having been omitted from the simple model in the earlier part of the chapter. Some of the ideas concerning it will come up again in the second part of the next chapter.

4

LONG-RUN MARGINAL COST

Long-run marginal cost, in present worth terms, is simply the present worth of all system costs as they will be with the increment in load which is to be costed, less what they would be without that increment. Since the load is growing from year to year and the expansion of the generating and transmission system is planned accordingly, an increment of load must be regarded as a marginal addition to the aggregate load growth which is already expected. As it is not convenient to cost an increment which is itself growing through time, we shall consider a marginal addition to the anticipated load in each future year which is the same for each future year.

In principle the load increment must be specified for each of the 8,760 hours of the year. (In practice it suffices to specify it for groups of hours or for a few representative days of the year.) Thus, to give a simple example, a load increment could be 70 kW in all winter working days from 8.00 to 17.00, and 25 kW the rest of the year. If the probability of the peak occurring between 8.00 and 17.00 on a winter working day is 0·9, the mean expected addition to peak demand is $0·9 \times 70$ kW $+ 0·1 \times 25$ kW $= 65·5$ kW. In what follows we allow in this manner for uncertainty concerning the *timing* of the peak, but will continue to postpone consideration of all other kinds of uncertainty to a later chapter. The level of the peak is taken to be known and to be increased by 65·5 kW in each year above the level it would then have in the absence of the specified load increment.

What this means is that a load increment is described both by a vector dU, specifying the incremental kWh taken in each of the 8,760 hours of the year, and by its mean expected addition to peak demand, denoted dP. If the predicted time of system peak is changing over the years, the dP corresponding to a given dU may also change, but we shall assume that it does not in order to keep the exposition simple.

The interest of long-run marginal cost stems from its relevance to tariffs. The size of load increment which needs to be costed is therefore the load of a single customer or a variation in the load of an individual existing consumer, since tariffs affect each consumer individually. Except for a very small number of large industrial consumers, this means that the size of load increment is microscopically small in relation to the total load. A purely marginal approach is thus required.*

A SINGLE NODE

We first examine the long-run marginal cost of generation only, neglecting transmission. Distribution costs are also ignored since we are continuing to measure the load at the point where current is fed into the distribution network. Distribution costs are examined in a later chapter.

The increment in system costs from a given load increment has three components. A minor one is the increase in what were termed system overheads in the chapter on the optimum plant mix. Since headquarters staff, research expenditure and other such items grow with overall system size, they have a marginal component. Whether they are kW or kWh related is not easy to determine, but some allowance must be made for them in costing a load increment.

Leaving this minor marginal cost component on one side, we have to consider two items:

dU. The extra kWh of load in each hour has to be multiplied by marginal running cost in that hour, m, giving a total in present worth terms of

$$PW[\mathrm{d}\,Um]$$

* The fact that capacity extensions are not divisible creates no problems since the cost per kW of an indivisible extension can be regarded as the average level of marginal cost over a range of output. In a large and expanding system, indivisibilities can be neglected. Neither, in the present context, is any problem created by the fact that a shift from non-marginal to marginal cost pricing may cause a large load increment or decrement. This large increment or decrement may alter marginal cost as defined here, but this does not impugn the relevance of the definition given.

dP. The mean expected addition to peak demand requires an equal addition to available capacity* at a cost of:

$$dP\ (C-PW[(m-r)a]+PW[F])$$

If the load increment is costed sufficiently far in advance for the optimal plant mix to be preserved, this second term will be the same for all types of plant. If the optimal plant mix is not preserved, this term will be a weighted average for the different types of plant, the weights corresponding to their respective contributions to dP.

TOTAL COSTS AND CAPITAL COSTS

Consider a load increment where dP is zero. It is evident that since this load increment requires no increase in capacity, its long-run marginal cost consists only of marginal running costs, at least in a system where the optimal plant mix is retained. This is not to say, however, that it will involve no extra capital expenditure; on the contrary, it may well do so.

In order to elucidate this point let us revert to the simple steady-growth model of the last chapter and concentrate on the gas-turbine/conventional plant mix, with the aid of the accompanying diagram. The load increment is shown by moving upwards part of the load duration curve to the dashed line. In order to make the diagram clear, a large increment is shown, but we are really concerned only with a very small one. It can be seen that the optimal plant mix to meet the load including the increment differs from the optimal plant mix to meet the load without the increment. The former involves dG less available capacity in the form of gas turbines and dG more in the shape of conventional plant. The net effect of this difference in mix upon annual system costs is:†

$$dG\,[A\{C_c\}-(r_g-r_c)H_g+F_c-A\{C_g\}-F_g]$$

With an optimal plant mix this expression is zero; i.e. the incre-

---

* Actually it would require a larger addition to keep down the risk of supply interruptions but this involves the problem of risk which is postponed to the next chapter. It suffices here to note that the addition to *installed* capacity will exceed the mean expected addition to peak demand by a reserve ratio (currently 17% in England and Wales).

† It will be recalled that gas turbines produce no fuel savings in this simple model.

mental load involves no marginal costs other than dUm. But it clearly does involve greater capital expenditure since:

$$C_c > C_g$$

This difference, with an optimal plant mix, is exactly compensated by the present worth of the difference in fixed other works costs and by the fuel savings which the extra conventional capacity will procure.

As a curiosum, it may be noted that if non-optimization of the plant mix took the special form of setting H_g, the maximum annual

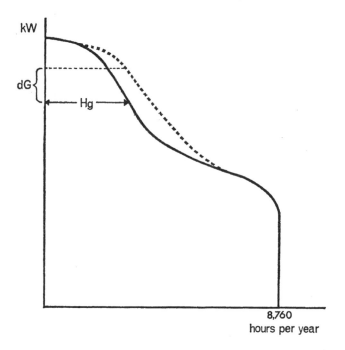

number of hours running of gas turbines, *below* its optimal level (i.e. having relatively too few gas turbines) the above expression would be positive. In other words, non-optimization of this sort would mean that the long-run marginal cost of a load increment would include not only the dP and dU terms, but also a term proportional to the increase in the load at H_g hours. This might be called incremental shoulder cost, since it involves the shoulders of the peak.

A similar argument applies to the conventional/nuclear mix. Thus a rise in the portion of the load duration curve at duration

$H_g + H_c$ would shift the optimal mix in favour of nuclear plant and raise total capital expenditure even though the total effect on the present worth of all system costs would still be only $PW[dUm]$.

If there were non-optimization in the special form of keeping $H_g + H_c$ too high (i.e. having relatively too little nuclear capacity) a rise in the load at duration $H_g + H_c$ would raise the present worth of all system costs by less than $PW[dUm]$. We could then speak of negative incremental base-load costs, since such a change in the load curve involves a rise in base-load.

INCREMENTAL COST WITH TWO NODES

We now enquire into the implications of the optimality conditions for the structure of long-term marginal costs when there are two nodes, thus bringing transmission costs into the picture.

An extra peak demand of dP at either node requires an equal increment of available capacity there or at the other node, assuming that their peaks coincide. The choice between the different ways of securing this increment must, at the margin, be indifferent, for if it were not the existing plant and location mix would be non-optimal. In other words the substitution of 1 extra kW of available capacity of one type for 1 kW less of another must leave the present worth of total system costs unaffected. This means that the cost of an extra kW of available capacity must be the same however it is achieved.

In considering this consequence of optimization it must be recalled that we have excluded from the analysis all distribution costs. Nodes are both load and generation centres. Hence we are implicitly supposing that extra generating capacity can be installed at the place where the load is located. This being the case, one possible way of meeting dP would always be to increase available gas turbine capacity at that point by dP. Even in the absence of complete optimization, the increment in the present worth of system costs resulting from this must be very much the same at all locations. The reason is that (local site differences apart) the capital and fixed other works costs, C_g and F_g, of gas turbines are the same everywhere and that since they have high running costs, their fuel savings must be small, so that *differences* in their fuel savings as between different nodes cannot be large.

Once again, equality of the incremental present worth of system costs does not imply equality of initial capital costs. An expansion of conventional thermal generating capacity requires much greater capital expenditure than an equal increment in available gas turbine capacity. But an optimal mix at each node implies that this difference is just offset by an opposite difference in fixed other works costs and fuel savings.

We now turn to the cost of generating the increment in kWh, dU. This obviously can differ from one node to another. At any hour the marginal cost of generating an extra kWh will differ between any linked pair of nodes on account of incremental transmission losses (α) or a transmission constraint ($\varepsilon > 0$).

Given that available capacity is increased by dP, no further capital expenditure is physically necessary in order to meet dU. On the other hand, some change in the system may become economically desirable. For example an increase in the load at a high fuel cost node i, dU^i, at times when i is importing from x, will make

$$C_t < PW[m^i\beta + dQ^x\varepsilon(1-\alpha)]$$

so justifying an increase in transmission capacity (or alternatively an increase in the share of nuclear capacity located at i). But this extra capital expenditure will only be justified if it enables the total load to be met at a lower present worth of all costs. Thus

$$PW[dU^i m^i] = PW\left[dU^i\left(\frac{m^x}{1-\alpha}+\varepsilon\right)\right]$$

is an upper limit to the increment in the present worth of system costs of meeting dU^i. At the margin, i.e. for a load increment which is very small in relation to the size of the system, optimization will secure equality.

The introduction of space and transmission costs into the analysis thus introduces geographical differences into the cost of dU but not of dP in an integrated system. It is true that the extra capital expenditure required by the load increment may differ from node to node but such differences will, in an optimized system, be exactly offset by differences in the present worth of fixed other works costs and fuel savings. But so long as attention is confined to small load increments and centred on the present worth of all system costs, geographical differences in the long-run marginal cost of meeting a load increment of $dP + dU$ are confined to geographical differences in $PW[dUm]$. These differences cannot be large.

VARIATIONS IN PLANT AVAILABILITY

A complication which has been skimmed over so far is that the availability of plant at peak times is different from its availability at other times. Leaving aside stochastic fluctuations in forced outages, this is because planned outages for maintenance are naturally concentrated into the months of the year when load is low. Thus if forced outages have a mean level of 10%, total outages will exceed this level by as little as possible over the peak months and be considerably greater at other times.

This has two consequences which deserve to be noted. The first is that the plant which is available will be changing throughout the year. Consequently marginal generating cost at any given load level will also be changing through the year. This clearly has to be allowed for. Thus the calculation of fuel savings requires that assumptions be made about the pattern of planned outages.

The second point is that if availability falls off as fast as load from the peak to the off-peak months, an increase in capacity may be required to meet a load increment even if that increment has no peak component. This extra capacity might take the form of increased transmission capacity. Thus, to put it in general terms, the dP component of the long-run marginal cost of a load increment must be measured with reference to the margin between load and capacity. At times of low load when a lot of generating plant is out for maintenance, the size of this margin may depend on transmission limitations. Hence it is conceivable that a load increment extending over the whole year may require additional transmission capacity in order to allow greater import into nodes where—for a period of the year—much of the generating capacity is not available.

THE CALCULATION OF LONG-RUN MARGINAL COST

In order to calculate the present worth of fuel savings and of the marginal generating cost of dU, it is in principle necessary to know m for each of the 8,760 hours of the next thirty or so years. This depends on the evolution through time of plant installed, the configuration of the transmission system, plant availability, security constraints and fuel costs. In practice, as pointed out earlier, a practical approximation can be achieved by simulating the operation of the system for a few representative days of the year for a limited number of years. Obviously the answers obtained will be no better than the assumptions and these must be simplified and are subject to a thousand and one uncertainties. What is important, however, is that the estimation of long-run marginal cost is consistent with the planning of capital expenditure. Both require the same calculations. Tariffs are fixed and capital expenditure decided only for a period of years; assumptions for later years are required but can subsequently be changed. Thus the choice of action is contingent upon assumptions which will probably turn out to be wrong in some respects. But sensitivity analysis may well demonstrate that neither long-run marginal cost nor the optimal capital programme are very sensitive to these assumptions. Furthermore, these assumptions are relevant whether one likes it or not, so explicit formulation of them is preferable to fudging.

MARGINAL AND TOTAL COSTS

The system existing at any moment of time is inherited from the past. Hence its annual capital charges and (given fuel prices and the costs of other current inputs) the cost of maintaining it and running it to meet a given load depend on the history of the growth of its capacity. If technological progress has outpaced inflation and if the load has grown fast, so that a high proportion of existing capacity was installed recently, total accounting costs will be low. If, conversely, inflation has outpaced technical progress, a high proportion of new capacity will mean high costs.

Scale economies abound. Some of them, which depend on system size, have already been discussed. Others occur at the level of the individual plant, e.g. a 500 MW conventional generating set costs less per kW to build than a 300 MW set. The extent to which advantage can be taken of these economies depends partly on the size of the whole system and partly upon the annual growth in capacity. These inter-relationships will not be examined here, since they are extremely complicated. But it does seem worth while showing that in a system without history and without any economies of scale the textbook relationship between long-run marginal and average costs holds good. Such a system was analysed in the previous chapter. The assumptions made there, it will be recalled, were uniform growth, no technical progress and the existence of only three types of plant.

51

Let us consider such a system in a particular year with a total available capacity of: G, gas turbines; C, conventional plant; N, nuclear plant; meeting a peak load of $G+C+N$. If we take a load increment with $dP = 1$ kW and dU such that the load in each of the 8,760 hours in the year is increased in the same proportion, namely $1/(G+C+N)$, we are looking at a load increment of exactly the same shape as the total load curve. Now the absence of history and of scale economies means that the long-run marginal cost of this increment (expressed as an annuity) is exactly $1/(G+C+N)$ of the total system cost (also expressed in annual terms). This is

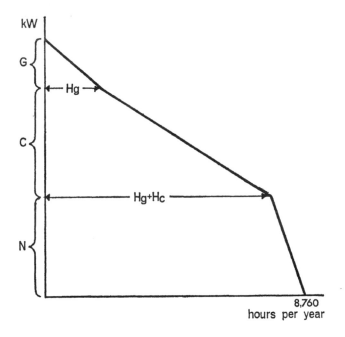

obvious, by virtue of the special assumptions. On the other hand, the incremental annual cost of an extra available kW of capacity is only:

$$A\{C_g\}+F_g$$

which is much less than the annual cost per kW of all available capacity, since conventional and nuclear plant are more expensive. This is just offset by the excess of marginal running cost over average running costs. But in case it is not intuitively apparent that this is so, a proof follows:

The procedure is to show that total annuitized cost divided by $1/(G+C+N)$ equals long-run marginal cost, expressed in annual terms, for the special case where each of the three segments of the load duration curve (H_g, H_c and $8,760-H_g-H_c$) is linear as illustrated in the accompanying diagram. Total cost consists of the sum of the annuitized capital cost per available kW multiplied by available capacity, for each of the three types of plant:

$$GA\{C_g\}+CA\{C_c\}+NA\{C_n\}$$

plus a similar sum for fixed other works costs:

$$GF_g+CF_c+NF_n$$

plus total running costs, equal, for each type of plant, to the product of kWh generated in the year and generating cost per kWh:

$$G\frac{H_g}{2}r_g+C\frac{H_g+(H_g+H_c)}{2}r_c+N\frac{(H_g+H_c)+8,760}{2}r_n$$

Adding up these components, substituting according to the optimality conditions:

$$A\{C_g\}+F_g = A\{C_c\}-(r_g-r_c)H_g+F_c =$$
$$A\{C_n\}-(r_g-r_n)H_g-(r_c-r_n)H_c+F_n$$

and re-arranging gives:

$$(G+C+N)(A\{C_g\}+Fg)+r_gH_g\left(\frac{G}{2}+C+N\right)+$$
$$r_cH_c\left(\frac{C}{2}+N\right)+r_n(8,760-H_g-H_c)\frac{N}{2}$$

as an alternative expression for total annual system costs. Divide this through by $1/(G+C+N)$ and we get long-run marginal cost since:

$A\{C_g\}+F_g$ is the incremental cost of 1 extra kW of available capacity;

$H_g\dfrac{\dfrac{G}{2}+C+N}{G+C+N}$ is the total increase in load during the hours when marginal generating cost is r_g;

$H_c\dfrac{\dfrac{C}{2}+N}{G+C+N}$ is the total increase in load during the hours when marginal generating cost is r_c;

$$(8,760 - H_g - H_c)\frac{\dfrac{N}{2}}{G+C+N}$$

is the total increase in load during the hours when marginal generating cost is r_n.

Since non-linearity of the three segments of the load duration curve merely affects running costs, altering total and incremental costs in the same proportion, this result holds even when the load duration curve does not consist of three linear segments.

If, therefore, system overheads are all kW or unit related too and can accordingly be included in the $A\{C\}$, F or r terms, the above argument shows that in the absence of scale economies and history, a generating system which sold electricity at long-run marginal cost would exactly cover its total costs.

THE TIME PATTERN OF MARGINAL COST

So far in this chapter, long-run marginal cost has been expressed as a present worth or as a constant annuity. What is really involved, however, is an infinite time stream of incremental system costs, including capital expenditure followed (say) every thirty years later by replacement capital expenditure, and administrative, maintenance and running costs every year for ever.

If we 'levelize' this time stream, i.e. find the constant perpetual annuity with the same present worth, it is tempting to divide the annuity by the annual kWh in the given permanent load increment and to call the result the long-run marginal cost of that increment. But in the context of relating tariffs to long-run marginal cost this would surely be relevant only if what was being examined was a *long-term contract* for a permanent load increment of the given shape. In other words, what the calculation gives is the price which, according to marginal cost principles, should be charged year after year for such an increment if the price is not to change (apart from inflation).

This does not seem to be a very useful concept for two reasons. First, electricity is rarely sold on such long-term contracts. Second, there is something arbitrary in taking a constant annuity when in fact it is generally expected that technical progress will cause costs to fall (inflation apart). There is an infinite number of time streams of pence per unit, each of which, when multiplied by the annual number of kWh in the load increment, has a present worth equal to that of the increment in system costs. The constant time stream is just the one which happens to be arithmetically most convenient.

The right way to isolate the share of the increment in system costs caused by a given permanent load increment which can be attributed to a particular year, year n, seems to be to compare:

(i) the present worth of the increment of system costs resulting from a permanent load increment starting at the beginning of year n; with

(ii) the present worth of the increment of system costs resulting from the same permanent load increment starting at the beginning of year $n+1$.

The excess of (i) over (ii) is then the present worth of the long-run marginal cost in year n of the given load increment.

We can now go back to the ingredients of long-run marginal cost as spelled out in an earlier section of this chapter and elaborate the new definition. In doing so it will be assumed that items (*b*) and (*c*) below are the same per kW of extra capacity commissioned in time for year n as for year $n+1$.

(*a*) Capital cost (including interest during construction) of dP kW of available generating capacity and associated transmission capacity ready for year n minus the corresponding figure for year $n+1$ discounted back to year n, plus the excess of the present worth of the cost of the infinite chain of replacements in the first case over the second case

(*b*) incremental administrative costs, rates, etc. in year n

(*c*) incremental manning and maintenance costs in year n

(*d*) the incremental generating costs (nearly all fuel) of generating the load increment dU in year n

(*e*) minus the difference in the present worth in year n of fuel savings with dP kW of additional capacity available for year n and for year $n+1$.

In the absence of all future technical progress, i.e. if the capital and running costs of new plant ready for year n were expected to be the same as for the following year, those of the above items which are most complicated become simple:

(*a*) is then simply the annuitized cost of the extra capacity

(*e*) is simply the fuel savings from the extra capacity in year n

(*d*), however, will still fall through time since new plant is cheaper to run than existing plant and will constitute a larger and larger part of the system as it expands. In other words, past technical progress still affects things even if future technical progress is ignored.

But if, as should be done, allowance is made for future technical progress, the result is to *raise* long-run marginal cost. The paradox is due to the fact that providing new plant for year *n* rather than for year *n*+1 involves the sacrifice of one year's technical progress and this is a cost. Let us spell this out in detail for (*a*) and (*e*). Take (*a*) first and let:

K = the capital cost of dP kW of available capacity ready for year *n*

p = the annual reduction in K, i.e. the rate of technical progress

r = the discount rate

Then we have to subtract the present worth of:

$$\frac{K(1-p)}{(1+r)}+\frac{K(1-p)^{31}}{(1+r)^{33}}+\frac{K(1-p)^{61}}{(1+r)^{61}}\cdots$$

from

$$K+\frac{K(1-p)^{30}}{(1+r)^{30}}+\frac{K(1-p)^{60}}{(1+r)^{60}}\cdots$$

assuming a thirty-year expected economic life. If $p = 0$ this equals the annual value of K annuitized at r over thirty years; taking $p>0$ is clearly equivalent to raising the discount rate and this gives a higher figure.

Now consider item (*e*), fuel savings. Technical progress, which makes the running cost of new plant ready for year *n*+1 lower than that of new plant ready for year *n*, will reduce the excess of the present value of the fuel savings provided by the latter over the present value of that provided by the former. Since this excess enters into the long-run marginal cost for year *n* as a negative item, we again find that taking technical progress into account raises long-run marginal cost.

DEPRECIATION, INTEREST AND MARGINAL COST

It may help in grasping this point if it is stated in more familiar terms. The anticipation of future technical progress implies the progressive obsolescence of the new plant now being installed. This suggests that the depreciation of the new plant should not be spread evenly over its life, but should be concentrated on its early years. Thus if we think of costs in terms of running costs plus annual capital charges, they will diminish through time.

All this is just to say with Boiteux that 'the language of amortization thus appears as equivalent to the language of discounting'.* But it is the economist's language of amortization, not the accountant's, which is involved. To be able to use it one should, as Boiteux adds, 'have clear appreciation of the manner in which amortization would be determined if one knew, today, everything that is needed to calculate it with precision'. Having spelled out what is needed, the ingredients of long-run marginal cost given in the previous section of this chapter can now be set out in an alternative form as follows. As always in this book, inflation is disregarded, i.e. the calculations are made for a constant general price level.

(a_1) Interest during the year on the capital cost (including interest during construction) of the increment in capacity less the amount depreciated to date

(a_2) depreciation during the year on the increment in capacity measured as the difference between its residual value to the system at the beginning and end of the year

(b) incremental administrative costs, rates, etc. during the year

(c) incremental manning and maintenance costs during the year

(d) the incremental generating costs of generating the load increment

(e) minus the year's fuel savings from the increment in capacity.†

The ingredient in this formulation which gathers up all the

---

* M. Boiteux: 'The Role of Amortization in Investment Planning', *International Economic Papers*, vol. 10, pp. 161–2, translated from *Revue de Recherche Opérationelle*, 1957.

† In order to check that this formulation agrees with the formulation of the previous section, consider once again the special case where no future technical progress is allowed for ($p = 0$). We saw then that in this special case:

(a) was the annuitized cost of the extra capacity

(e) was the fuel savings from the extra capacity.

The latter is thus obviously the same under either formulation, so we have only to explain why (a) = (a_1)+(a_2). Obviously (a_1) falls from year to year by r times depreciation of the previous year, r being the interest rate. But depreciation rises by this amount each year because the loss of residual value consists of the approach of the time when replacement expenditure will be required and each year's approach raises the present worth of that expenditure by r.

expectations and calculations relating to the whole of the future lying beyond the year in question is (a_2), the year's reduction in the residual value to the system of the specified increment in capacity. This is clear if we recall that the definition of the residual value to the system at any point of time of this increment is:

(i) the present worth of system costs from then on as they would be if that increment had not then been inherited from the past; minus

(ii) the present worth of system costs from then on as they will be.

Residual value is the shadow value of the existing capacity constraint in a programming formulation of the investment optimization problem. This can be illustrated diagrammatically with the aid of the curve showing those combinations of running cost and annuitized capital cost plus fixed other works costs which are equally attractive at the margin in the case of a new plant programme (see page 18). Point z there represented existing plant. Draw a horizontal line through point z from the point on the vertical axis representing the running cost of the plant to the isocost curve. Subtract the fixed other works costs of this existing plant from this line and the remainder measures what the annuitized residual value of the existing plant would be if it had the life expectancy of similar but new plant. The difference between this residual value, in present worth terms, in two successive years will thus equal the depreciation such plant would suffer and this will differ only trivially from that to be imputed to the existing plant in its actual state.

CONCLUSIONS

Long-run marginal cost having been defined in terms of a permanent load increment it has been shown to consist of three components:

(i) system overheads
(ii) running cost, related to dU
(iii) capacity costs.

Provided that the permanent load increment starts sufficiently far in the future for capital programmes to be fully and optimally adjusted to it, it has been shown that (ii) but not (iii) will vary geographically in an integrated system where peak output occurs at the same time in different regions. An important distinction has been drawn between long-run marginal capacity costs which are entirely peak-related, on the one hand, and extra capital expenditure on the other. It was noted that the dP component of a load increment must be measured in relation to the margin between load and capacity. It coincides with the simple notion of peak demand only in systems where the seasonal modulation of demand is large in relation to the seasonal fluctuation in maintenance and hence in plant availability.

An *a priori* statement about the relationship between long-run marginal cost and total cost is possible only if the system lacks history—i.e. in practically no real system. Thus simple statements of the sort that, since the industry is one of increasing returns, marginal cost pricing would involve a deficit, are irrelevant.

The same fact of technical progress makes long-run marginal cost a much more complicated concept than any economics textbook allows. There is no escaping an element of judgement in its calculation. It has been shown that this judgement is required concerning a large number of variables but that it can also be packed into one portmanteau guesstimate about depreciation.

Finally, it cannot be emphasized too strongly that any estimate of long-term marginal cost has no significance *in abstracto* but only in relation to a specified load increment. There are as many marginal costs as there are conceivable load increments.

5

RISK AND UNCERTAINTY IN GENERATION

The analysis of the preceding chapters has treated all the variables as single-valued. It has, in other words, been supposed that the estimates made of load and of the various cost items are either single best estimates or else mean expectations and that these are then appropriate to use in the calculations as though they were known with certainty. Despite some references to the unknowability of the future and despite the point that the timing of the peak was on one occasion treated as stochastic, the discussion has thus left a major problem on one side. It is now time to look at this problem.

The main elements of risk and uncertainty which deserve attention can be grouped as follows:

(*a*) LOAD. This really involves two separate problems:

    (i) Forecasting the load under the assumption of specified standard weather conditions. The errors in past forecasts give some guide to the range of error that may be expected in the future. But if forecasting techniques are improving and if the past errors have not been distributed round zero (i.e. if past forecasts have shown some systematic bias) it is not possible to calculate any relevant measure of the range of error to be expected in future forecasts—which are intended to have no upward or downward bias. Retrospective forecasts of past loads using current forecasting techniques would provide such a measure only if the techniques were fairly mechanical and if the statistics they require were all available for earlier years.

    (ii) Allowing for deviations of actual weather from standard weather conditions. Since past experience is a reasonable guide to the future in this case, frequency distributions of the difference between actual and 'standard weather' load can be calculated and used if, as we assume, enough is known about the response of load to weather deviations.*

(b) PLANT AVAILABILITY. The uncertainty relates both to delays in the construction of new plant and to unplanned outages. Given statistics about the frequency and duration of such outages for each type of existing plant, engineers can make estimates concerning types of plant not yet built. Given the mix of the system, and given an allowance for construction delays, calculations can then be made, either by probability calculus or by simulation, of the overall plant availability in the system and of its variance.

(c) FUTURE RELATIVE PRICES AND TECHNOLOGY. It has been supposed, for example, that the capital cost per kW and the running cost per kWh of a nuclear plant to be built in fifteen years' time can be estimated. Even though this does not require a prediction of the amount of inflation, since it is only relative prices that matter for socially optimal investment decisions, such an estimate can only be an informed guess.

We thus see that weather risk and availability risk can be dealt with in terms of frequency distributions; they can be the subject of probability calculations of more or less sophistication according to the availability of data. The pure forecasting risk can be treated in a similar manner by the arbitrary but useful expedient of attributing a certain variance to forecasting errors in the light of experience and hope. The uncertainty about prices and technology, however, can be expressed in probability terms only if engineering planners are prepared to express their guesses as subjective probability distributions, something which seems unlikely.

Decision theory has no solution to the problem posed by non-probabilizable uncertainty. Thus in the preparation of a background plan it seems that a set of plausible guesses about the future must be considered. It may well turn out that differences between the optimal background plans corresponding to some of the different guesses will not have a large effect on the optimal shape of the next annual capital programme. If so, the final choice of the latter

---

* In a system with hydro-electric plant there is also a quite different risk connected with weather, namely variations in the availability of water. This topic is not examined here.

is not too difficult. Otherwise pure judgement is required, aided by information about the optimal capital programme corresponding to each sub-set of guesses.

Having dutifully mentioned uncertainty about prices and technology, this chapter concentrates upon the forecasting, weather and availability risks. It is assumed that the forecasting risk has been judged and expressed in probability terms so that probability calculus or simulation can be applied to all three types of risk either singly or in combination. Three of the topics discussed in earlier chapters now need to be looked at again in probability terms. In all of them it is assumed that the aim now is to minimize the *mean expectation* of the present worth of system costs, subject to the relevant constraints in each case.

The first, which is no more than mentioned here, is the hour to hour operation of the system, i.e. economic despatching. From what was said in the first chapter it is clear that decisions concerning hot standby, spinning reserve and out of merit order operation motivated by the need for security are related to the possibility that the load will not move as expected or that there will be forced transmission or generating plant outages. Operating rules of thumb have to be developed about the extent and manner in which the risk of supply interruptions will be limited—for example a rule that spinning reserve at any time must not be less that the capacity of the largest set then operating. The rationale of such rules is in principle to be sought in terms of functional relations between security (i.e. the risk of failure to meet the load) and cost.

ESTIMATING FUTURE RUNNING COSTS

Calculations concerned with optimizing the plant mix or the estimation of long-run marginal cost constitute the second topic where risk plays a part. These calculations involve simulating the future operation of a given system to meet an assumed load. We now have to bring in the complication that weather and forecasting risks make the load assumption a frequency distribution for each period, instead of a single figure and that the same is the case with the assumption as to what plant will be available during that period.

Consider the problem of calculating m, marginal generating cost, for a single hour. Suppose that the risk is limited to the very simple case where the load in that hour will be either OL_1 or OL_2 with equal probability. The mean expectation is then $m\ (= L_1m_2 + L_2m_2)/2$ and it can be seen from the accompanying diagram that because of the curvature of the marginal cost curve this is greater than Lm, marginal cost at the mean expected load. A similar argument for availability risk produces a similar result. Hence, to generalize, the mean expectation of m is not the same as the m corresponding to the combination of mean expected availability and mean expected load. The sort of single-valued calculations of earlier chapters using mean expected values for availabilities and loads will thus give systematically biased estimates of the mean expectation of m.

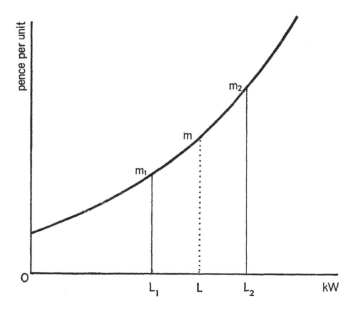

No general statement is possible about the best way to handle this complication. Data limitations, computational capacity, a desire for simplicity and how much accuracy is required are all relevant. One obvious path to follow is the calculation of m for each of a limited number of sample combinations of load and availability, giving each combination an appropriate weight, so that a mean expected value of m can be obtained. Alternatively, m might be calculated for the mean expected load and availability and then corrected by a factor dependent on the variance in the load and availability distributions and the curvature of the marginal cost curve in its relevant stretch.

GENERATING RESERVE REQUIREMENTS

The third topic where risk introduces complications hitherto ignored is that of the required generating capacity. We have hitherto assumed that installed capacity less mean expected outage at time of peak is set equal to the mean expectation of peak demand under standard weather conditions for the year in question. If this were done:

> demand under standard weather conditions in excess of the forecast
> weather worse than standard conditions
> availability below the mean expectation

singly or in combination would result in an excess of demand over availability at peak at 50% of the times of peak demand, on average. On such occasions the load would have to be reduced, either by voltage reductions or by load-shedding.

An excess of mean available capacity over the mean expectation of demand under standard weather conditions will obviously reduce the risk of load reduction. In order to reduce that risk it is thus necessary that the total margin of installed generating capacity over the mean expectation of peak load under standard weather conditions shall exceed the mean expectation of non-availability at peak. This plant margin adds to system costs. Hence the problem arises of determining how much the risk of load reductions can be diminished for a given addition to costs or, conversely, the extra cost of a given diminution in the risk. Once this is known the choice of the appropriate level of risk to aim at will be illuminated.

Since the net addition to system costs of an increment in capacity has already been discussed, what needs to be examined here is the relationship between the level of installed capacity and the risk of load reductions.

The hours of the year when this risk is large enough to be worth bothering about may be called the 'critical period'. In a system where seasonal variations in the load are large relative to the seasonal variation which can be achieved in planned outages for maintenance, the critical period will be confined to the peak and near-peak hours of the days of high demand in the season when the load is highest. At the opposite extreme, when seasonal variations in the load are small enough to be more or less matched by seasonal variations in

planned maintenance, the critical period will encompass the peak
and near-peak hours of most working days in the year.*

Let us consider one hour, or a set of similar hours, within the
critical period. Given the plant existing, excluding that undergoing planned maintenance, a probability distribution of available
capacity can (in principle) be calculated, given frequency distributions
of the occurrence and duration of forced outages for each of the
types of plant existing. This will reflect not only the probability of
forced outages occurring in the hour or hours concerned, but also
the probability of forced outages earlier on with a duration extending to this hour. This can be expressed by saying that it is not

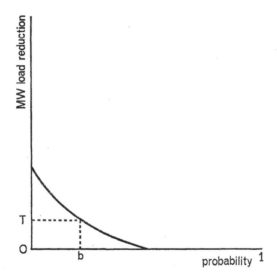

just the risk of a forced outage occurring which matters, but rather
the risk that plant will not be available owing to forced outage.

Given a frequency distribution of weather conditions, an estimate
of the weather response of the load, a forecast load under normal
weather conditions and a judgement of the variance of forecasting
errors, a probability distribution of load in the hour or set of similar
hours concerned can also (in principle) be calculated.

The two probability distributions can now be combined. Thus the

* In a hydro-electric system the risk of water shortage may give rise to a
critical period even at times when capacity is high in relation to demand. In
such cases, the risk may have a cumulative element due to gradual exhaustion of
stored water.

probability of load reduction of y MW in each hour concerned is the sum of:

(probability of load j) × (probability of available capacity k) for all combinations of j and k where $j-k = y$.

If this is done for all values of y the end-result can be expressed as a cumulative probability curve for the hour or hours concerned as shown in the accompanying diagram. This shows, for example, that the probability of a loss of load of OT or more per hour is Ob. Thus the area under the curve measures the mean expectation of load reduction per hour.

Data limitations and the complexity of the calculations may make impossible a complete calculation along these lines for all the hours of the critical period. A whole range of approximations can be resorted to, however.* The end-product may be some measure of the risk of load reduction as a function, *inter alia,* of the level of installed capacity. This would show the reduction in risk (gain in security) obtainable by increasing the plant margin and hence could serve as a basis for decision on the size of margin to be adopted. More simply, the end-product may be the installed capacity required in order to meet a given forecast load at some previously established level of security.

INCREMENTAL CAPACITY COSTS AND RISK

Once a decision has been taken about the level of risk of load reduction which is to be incurred, the incremental generating capacity requirement of a load increment whose long-run marginal cost is to be ascertained can be estimated. It is the increase in installed capacity which is required to keep the risk of load reduction the same with the load increment as it would be without it. This can be ascertained by multiplying the maximum size of the load increment during the critical period by $(100+R)/100$ where R is the percentage gross plant margin required by the chosen risk level.

This makes little difference to the approach of the previous chapters which was expressed in terms of mean availability at peak.

* An example of an approximative graphic technique is furnished by G. S. Vassall and N. Tibbarts: 'An approach to the analysis of generating-capacity reserve requirements', *IEEE Transactions: Power Apparatus and Systems,* January 1965. The paper itself and the 'Discussion' illustrate the complexities of the problem.

The earlier discussion stands, subject to an appropriate modification in the terminology. In the expression for the net change in the present worth of all system costs arising from a 1 kW change in peak demand, namely:

$$A\{C\} - PW[(m-r)a)] + F$$

we now (*a*) measure this per *installed* kW, multiplying it by dP $(100 + R)/100$ in order to get marginal capacity costs, and (*b*) treat the coefficient *a* as being zero instead of unity not only when the new plant is out of merit but also when it is out for maintenance, planned or unplanned.

6

INVESTMENT AND COST STRUCTURE
IN DISTRIBUTION

THE CONTRAST WITH GENERATION AND
TRANSMISSION

Distribution is much less capable of analytical treatment than are generation and transmission. This is because the concept of overall system optimization really has no meaning in the case of distribution. While generation in one part of the country can be substituted for generation in another part, provided that there is sufficient transmission capacity, each part of the distribution system is specific to the consumers served by it. Again, while a new generating station can displace some generation by older plant with a higher running cost, new mains and transformers add to distribution capacity only in the particular locality where they are installed.

This accounts for a big difference in the way investment is planned. The total of new generating capacity and the plant mix are related to total load growth and to the characteristics of the whole generating system. Most investment in distribution, on the other hand, is nothing but the sum of a very large number of individual schemes, each determined either by the prospective growth of load in relation to distribution capacity in a particular locality or by a need to replace a particular unsafe or obsolete piece of equipment. A given requirement for extra capacity to meet, for example, the electricity requirements of a new factory will result in quite different capital expenditure in bringing the supply to the consumer's terminals according to the location of his factory in relation to existing supply lines, their voltage level and their present loading.

While the fact that each case is different excludes any overall optimization, it naturally remains true that investment decisions have an economic content. A new housing estate can be supplied with electricity in various ways, involving different layouts, cable sizes, substation sites and so on. The cheapest must be chosen, subject to constraints on voltage drop and security of supply. Again, if the capacity of the high voltage supply to a town has to be increased, the planning engineer will be able to design several alternative reinforcement schemes. The choice between them may not simply be a matter of cost minimization subject to security constraints, since one scheme may provide a larger increment of capacity than another, thus allowing for a larger number of years of future load growth. But the choice is still between a limited number of specified alternative schemes each designed for the particular circumstances of place and time.

All designs for schemes must use standardized components and conform to the design standards of the system of which they will form a part. Take transformers, for instance. It is sensible to use a limited range of sizes and types because of economies in purchasing, in spares and in training the labour force. Similarly with voltage levels. Even if a thorough investigation showed a non-standard voltage level to be ideal in a particular scheme considered in isolation, the theoretical gain to be had from using it will be outweighed by the extra trouble and cost of designing, installing and maintaining the lines and equipment.

Standards have to be revised from time to time and this naturally provides scope for more general analysis. Thus the range of transformer sizes to use, security standards and the choice of voltage levels in primary distribution are all topics for such analysis. Special investigations may thus be carried out to provide a basis for choosing the standards which are taken as given by the engineers designing particular schemes. An example would be an exercise to determine optimal voltage levels and transformer sizes as a function of load density in an imaginary geometrically uniform network. If the results proved fairly insensitive to variations in the values of the parameters, they could usefully be applied as a guide for actual systems.

All this shows that while engineering economic analysis can find plenty of scope in distribution, it does not provide information about cost structure as a by-product of a general optimization procedure. In contrast with generation and transmission, therefore, it is necessary to adopt an inductive approach to costs in the case of distribution.

SOME GENERAL FEATURES OF FIXED CAPITAL INVESTMENT

Fixed capital expenditure in distribution in Britain today can be very roughly broken down as follows:

New Business	30–35%
Reinforcement	50–55%
Other	15%

The latter consists mainly of non-operational land and buildings, vehicles and mobile plant.

Only part of the expenditure incurred under the New Business heading (which includes services and meters) can be simply allocated to particular new consumers. Some of it may relate to groups of consumers, e.g. in rural electrification, and even when an individual new consumer is connected, the extra capacity put in may jointly serve that consumer and reinforce the supply to other neighbouring consumers. Thus at least two thirds of capital expenditure on distribution is a joint cost. This is one important general feature.

Another general feature follows from the fact that, with the trivial exception of vehicles, mobile plant and tools, most of the capital assets used in distribution are long-lived. This means that when, as is the case in most countries, the size of the industry has been expanding continuously over many years, the amount of replacement expenditure will be only a small fraction of total expenditure. Thus suppose that all assets had a thirty-year service life, that new investment in each year was proportional to load growth and that the rate of load growth averages 8%. Then total investment now would consist of new investment proportional to this year's load growth plus replacement equal to the new investment of thirty years ago in the ratio 10:1. Thus 91% of the investment related to load growth, i.e. the categories New Business and Reinforcement, would be new investment and reinvestment would constitute only 9% of the total. This figure is no more than a guess but even so it does serve to make the general point that reinvestment is relatively unimportant. No direct estimate of reinvestment is possible, since like is not replaced with like.

A third general feature is that the capital costs of a distribution network normally fall as the load density (MVA per square mile) increases. This appears to be true not merely of a new network

70

built from scratch, where alternative schemes for meeting different sized loads in a given area are considered, but also when increments to an existing network are examined. In other words it seems that with a given technology and price level the capital cost per incremental kVA of adding to distribution capacity within a given area is generally a decreasing function of the existing capacity within that area. There may, in some cases, be discontinuities, however, as when the stage is reached when new extra high voltage infeeds have to be provided into the centre of a densely populated urban area.

These economies, which can scarcely be called 'scale economies' since they arise from an increase in load within a constant area, may arise partly because of scale economies at the level of individual items of plant and equipment. The cost per unit capacity of transformers falls as their capacity rises. The cost of digging a trench is much the same for a small low voltage main as for a large one. An increasing load density makes it possible to take advantage of these economies.

More important is that once an area is electrified an increase in the load in that area requires less than proportionate increase in the circuit mileage of mains. While a new housing estate will require new low voltage mains, growth in the load on an existing low voltage network will be met largely by increasing the number of infeeds from the next voltage level (most commonly 11 kV in the UK). Thus while the capacity of distribution transformers must keep pace with the load, the number of circuit miles of high voltage mains will rise more slowly as average distances fall with the increase in load density, and the number of circuit miles of low voltage mains will increase more slowly still.

All this can be roughly illustrated with figures relating to the London Electricity Board for the ten years ending March 1966.* The load rose by about three-quarters, while the number of 33 kV equivalent miles rose only by about a third, the number of 11 kV equivalent circuit miles by about a tenth and the number of low

* *Source:* Annual Report of the Board. I have used figures for London rather than for England and Wales as a whole in order to avoid the complication of rural electrification and because of the smaller relative importance of high voltage consumers in London. Nevertheless, the figures are extremely rough because I have treated ½ mile of 66 kV as equivalent to 1 mile of 33 kV mains and both ½ mile of 22 kV and 2 miles of 6·6 kV as equivalent to 1 mile of 11 kV mains. Furthermore much work was done during the period to replace nonstandard supplies.

voltage circuit miles by about a twentieth. The installed capacity of distribution transformers rose by more than the load, but a major reason for the excess was probably an improvement in security standards and the working-off of a backlog. Thus the pattern conforms to what was suggested above; the incremental capital costs of load growth fall as load density increases.*

A fourth general feature which deserves remark is that the incremental capital costs of load growth fall through time (inflation apart) not only because of increasing load density but also on account of technical progress. This is less dramatic than in the case of generation but has nonetheless been important.

THE CAPITAL COST OF MEETING A LOAD INCREMENT

We can now consider the capital cost of meeting a permanent load increment. This can be divided into the New Business cost and the Reinforcement cost, the former now being re-defined more narrowly than in the accounting classification as the cost of plant, cables, etc. which are wholly and exclusively provided in order to meet that load increment. This is normally negligible in the case of a growth in the load of existing consumers and thus mainly arises in respect of the connection of new consumers, the exception being a large increase in the load of a large individual consumer.

Consider, as one example, a new housing estate. Connecting the houses (or flats) involves the provision of a meter and a service to each dwelling, the laying down of a low voltage network to feed all the services, the provision of a number of transformer substations to feed the low voltage network and connection of these substations to the existing high voltage network. The cost per dwelling of all this naturally depends on a whole host of local factors, but in addition there are some more systematic factors which can be listed as follows:

(a) The after diversity maximum demand per dwelling (ADMD) which is provided for. This is the maximum demand in kW of the group of dwellings divided by the number of dwellings and is less than the average of the maximum demands because these

---

* Where increasing load density reflects the urbanization of previously rural areas an increase in the proportion of under-grounding and greater difficulty in obtaining substation sites will work in the opposite direction, however.

do not coincide. Cost rises with ADMD, but less than proportionately.

(b) The extent to which provision is made for future growth of demand. Provision may, for example, be made for an initial ADMD of 6 kW but the network may be designed so as to facilitate later reinforcement when ADMD rises above this level. Extra initial costs are thus incurred now in order to reduce potential reinforcement costs in the future. While such a decision must clearly rest on some sort of optimization calculation, it is not worth while making such a calculation on each occasion. Hence some general rule is adopted, to be tempered by judgement in each particular case.

(c) Undergrounding, low housing density and distance from the existing high voltage network all add to costs.

(d) Cost per dwelling, *ceteris paribus*, is generally lower in a large estate than in a small one.

As a second example, consider a new factory or an extension to an existing one. If the factory has a large power requirement the new supply may involve a 33 kV line and the installation of one or more transformers to step the voltage down to the level used in the factory. Some switchgear may also be required.

These two examples show the nature of the New Business cost. It is evident that this can differ widely from one case to another even when the two permanent load increments are electrically identical. This does not mean, however, that ascertainment of the cost is an expensive luxury, since it will automatically be obtained in every case as part of the process of planning and authorizing the necessary capital expenditure.

In order to examine the nature of the Reinforcement costs let us take up the housing estate example, supposing that the estate is connected to an existing 11 kV network which in turn is fed from a 33 kV network which is fed from the transmission grid. Reinforcement costs arise because the new housing estate load—the permanent load increment under consideration—adds to the load on the existing 11 kV and 33 kV networks. A growth in the load of an existing housing estate does this too. What is relevant here in both cases is the addition to the peak loads on these networks. If the 11 kV network and the 33 kV network peaks are at different times, the addition to their peak loads will be different. Thus 11 kV Reinforcement cost depends on the size of the housing estate load at the time of the 11 kV system peak, while 33 kV Reinforcement cost depends on its size at the time of the 33 kV system peak. Only if all the other loads served by these high voltage systems have the same time-shape as the housing estate load will its peak be the same as the addition to their peaks.

Now capacity is increased in steps; it cannot be increased continuously because mains and transformers are indivisible items. Hence they normally have an unused margin of capacity in relation

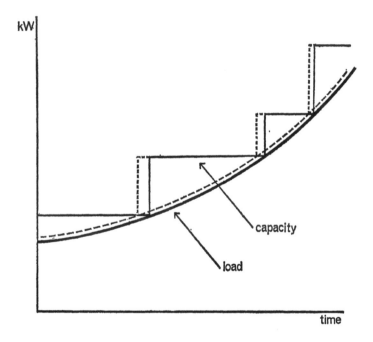

to their peak demand. The increase in this peak demand caused by the addition of the housing estate load will reduce this margin. If it is made negative, reinforcement of the high voltage network concerned is required immediately. If it is merely reduced, on the other hand, the date at which future reinforcement is required will be brought forward. This is because the environment is generally one of continued and continuing load growth which will in any case eventually fill up the margin of capacity and necessitate reinforcement.

The nature of Reinforcement costs is now clear. They are related

74

to the increase in peak demand on that part of the system considered and they consist of the bringing forward in time of reinforcement which would sooner or later have become necessary in any case. Formally speaking, Reinforcement costs at each voltage level of the existing network upstream from the housing estate equal (*a*) the present worth of all expenditure to infinity in expanding that network as it will be when brought forward in time, less (*b*) what it would be if it were not thus brought forward.

The diagram opposite may make the matter clear. The vertical axis shows the peak load on the existing high voltage system upstream from the new load, while time is measured along the horizontal axis. The solid curve shows the future growth of that peak load in the absence of the new load and the solid stepped line shows the appropriate growth of capacity. The dashed curve shows the growth of that peak load including the new load and the dashed stepped line shows the required growth of capacity corresponding to it. The Reinforcement cost of the new load is the present worth of the costs of achieving the dashed capacity-growth line minus the present worth of the costs of achieving the solid capacity-growth line. It consists both of an interest component and a technical progress component—an interesting parallel with our earlier analysis of generation capacity costs.

ESTIMATING REINFORCEMENT COST IN PRACTICE

This concept is obviously non-operational, if only because there is not perfect foresight. But the calculation of Reinforcement costs for each and every possible load increment would be required only if a separate tariff were to be quoted for each and every such increment. In practice, as will be explained in a later chapter, there have to be uniform tariffs for groups of consumers. Thus insofar as incremental Reinforcement costs are relevant to tariff making, what is wanted is an average of the incremental costs over a large number of load increments. This average can be expressed per kW of addition to peak demand on the existing system for each voltage level.

Imagine that 1,000 load increments are analysed in the impossibly perfect manner of the last diagram. Imagine that the resulting 1,000 diagrams are added vertically. All the steps in the capacity-growth lines will disappear in the new total diagram. The dashed capacity line now looks like a simple upward displacement of the solid capacity line. Thus the sum of 1,000 increments to peak load requires an addition to capacity now of *EF* MW. *EF* exceeds this sum by the average unused margin of capacity over peak load. Call this *r*. Then the Reinforcement cost per kW of increment in peak load is $1+r$ times the average capital cost per kW of capacity reinforcement being undertaken currently.

So we end up with a simple answer after all! On average, a permanent load increment whose size at the time of the peak load on the 11 kV system upstream is *Z* involves an 11 kV Reinforcement cost of $(1+r)Z$ times the cost per kW of 11 kV reinforcement. This is a

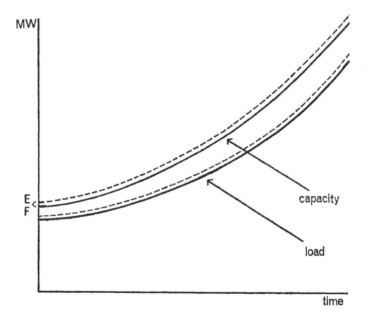

pretty obvious result, but at least we now know that it is completely consistent with the much more elaborate notion of incremental capacity cost which is necessary in thinking about generation and transmission.

The simplicity of the result does not, unfortunately, mean that it is easy to make the cost estimates because there is a wide variation from case to case. For example the cost per MVA of firm capacity of establishing new 33/11 kV or 33/6·6 kV new primary substations in one sample of seven such substations showed a range of three to one. (The reason for the highest cost was that particularly long lengths of underground 33 kV mains had to be laid.) Thus the costs need to be obtained as a statistical average of all cases rather than as an engineering cost estimate made for a representative or standard case.

CURRENT COSTS OF DISTRIBUTION

A breakdown of revenue expenditure in England and Wales during 1965/66 gives the following results.*

Running, repairing and maintaining the distribution network (buildings, transformers, mains, meters, etc.)	42%
Consumer Service	15%
Meter Reading, Billing and Account Collecting	12%
Training and Welfare	6%
Administration and General Charges and Expenses (including rents, postage, stationery, publicity, etc.)	25%

Nearly three-quarters of the total consists of wages, salaries, superannuation and training, i.e. of labour costs. This does not mean that distribution is especially labour intensive, of course, since these proportions relate only to current expenditure.

The cost of running, repairing and maintaining the distribution network depends on its size and nature, that is to say upon a whole host of characteristics, few of which are readily quantifiable. The cost of repairing and maintaining meters apart, to express this cost as a function of the number of consumers and the characteristics of their loads would be enormously difficult. Thus there is little hope of ascertaining the structure of these marginal costs in a way which is founded upon a quantitative representation of the distribution system.

Consumer service cost depends upon the number, types and

---

* *Source:* Annual Report of the Electricity Council, 1966. The following items have been excluded in order to consider only current distribution costs: Purchase and generation of electricity, Depreciation, Contribution to Electricity Council (which includes research expenditure and some publicity), costs allocated to appliance sales and contracting, repairs and maintenance of apparatus on hire and of public lighting, Changes of system of supply (consumers' appliances) and Rates. Printing, postage repairs and maintenance of offices, etc. have not been allocated but are all included with other General Expenses.

geographical dispersion of consumers and upon the standard of service given. Given these two latter factors and assuming that average and marginal costs coincide, a rough estimate of annual cost per consumer of each type could be constructed.

Meter reading, Billing and Account Collecting cost depends upon the number of consumers of each type, their geographical dispersion and upon the proportion of pre-payment (i.e. slot-meter) to credit consumers. Once again, a rough figure of cost per consumer of each type could be constructed, since there is no obvious reason why average and marginal costs should differ in the long run.* The cost of repairing and maintaining meters, included above under an earlier heading, could appropriately be added in here.

Finally, the cost of Training and Welfare, Administration and General Charges and Expenses, amounting to nearly a third of the total of current distribution costs, is mainly a joint cost. It may be that some parts of it can be meaningfully allocated to the other heads, but even so a large residue will remain. This big lump of overhead presumably grows along with the size of the distribution system, but it is anyone's guess what measures of size are relevant and whether it grows faster or slower.†

SUMMARY

It is abundantly clear that an attempt to ascertain the incremental distribution costs of a specified permanent load increment involves putting up figures which range from engineering cost estimates to nothing more than intelligent guessing. There are two circumstances which make this state of affairs supportable. The first is that average cost cannot be ascertained any less inaccurately—and that it may indeed, by allocating strictly unallocable costs, even give a less sensible answer. The other is that explicit estimates and guesses organized in an analytically sensible manner are better than blanket

---

* A simple regression of the 1965/66 Meter Reading, Billing and Account Collection costs of the twelve Area Boards against the number of their consumers of all kinds at March 1966 gave:

$$\text{Annual cost} = £64,657 + £1.006 \text{ per consumer}$$

with an r^2 of 0·926. The constant term accounts for only 8·25% and 3·3% of the relevant cost in the case of the Boards which spent least and most respectively. In view of the omission of other factors, this is small.

† See the end of the chapter for a discussion of the effect of distribution losses.

intuition of a final result, save for the rare case when extra-sensory perception comes to the aid of the tariff-maker.

In conclusion, let us list the items of incremental distribution cost, some or all of which will be relevant in considering a given permanent load increment.

(a) New Business capital cost in the narrow, allocable, sense. This is the only item which is automatically obtained for other, administrative purposes. It arises in the case of new consumers and large existing individual consumers such as a factory, but not in the case of growth in the existing load of small consumers.

(b) Reinforcement capital cost, the sum of one cost for each relevant voltage level. Each of these is the size of the load increment at the time of the peak on that part of the system, grossed up by the capacity margin and multiplied by an average cost per extra kW of capacity.

(c) The increase in the annual cost of running, repairing and maintaining the distribution system.

(d) The annual Consumer Service, Meter Reading, Billing and Account Collecting and meter repair and maintenance cost for the type of consumer in question (if the load increment comes from a new consumer).

(e) Some contribution to Training and Welfare, Administration and General Charges and Expenses (plus any taxes borne by the distribution undertaking).

Finally, the need to allow for distribution losses in costing the energy itself is explained below.

NOTE ON DISTRIBUTION LOSSES

The existence of distribution losses means that the cost per kWh supplied to consumers by the distribution system exceeds the cost per kWh fed into the distribution system. In order to examine this point in marginal terms, a simple theoretical exposition is required.

The losses K_{ti} in any hour (t) on any section (i) of a distribution system are the sum of:

F = iron losses, if the section includes transformers. No time subscript is necessary, since they are constant from hour to hour.

CL^2_{it} = copper losses; where C is a proportionality constant which depends on the physical properties and voltage level of the section, and L_{it} is the load on the ith section in the tth hour (i.e. consumer load served direct from that section + the demand of the following section).

Let there be n similar sections, each carrying an equal share of the load in the

t^{th} hour $L_t = \Sigma L_{tt}$. We now consider losses on all these similar sections, thus in effect speaking of total losses at a given level in the system.

These total losses in the $t^t{}_h$ hour, ΣK_{ti} are:

$$K_t = n\left[C\left(\frac{L_t}{n}\right)^2 + F \right]$$

i.e. n times the losses on one section.

Let L_p be load in the peak hour and assume that capacity is expanded *pari passu* with peak load so that:

$$n = EL_p$$

i.e. reinforcement proceeds by adding to the number of sections so that, on average, maximum demand on each section is preserved at $1/E$.

Substitution into the expression for total losses in the tth hour at the level of the system under consideration gives:

$$K_t = EL_p\left[C\left(\frac{L_t}{EL_p}\right)^2 + F \right]$$

$$= L_t^2 \frac{C}{EL_p} + EL_p F$$

so that total annual losses are, summing over $t = 1 \ldots p \ldots 8{,}760$

$$\Sigma K_t = \frac{C}{EL_p} \, \Sigma L_t^2 + 8{,}760 \;\; EL_p F$$

This gives marginal annual losses with respect to load in the peak hour as:

$$\frac{\partial \Sigma K_t}{\partial L_p} = \frac{C}{E} - \frac{C}{E} \cdot \underset{t \neq p}{\Sigma} L_t^2 \cdot \frac{1}{L^2} + 8{,}760 \; EF$$

The first term on the right-hand side is simply marginal copper loss in each peak hour. Since capacity is expanded *pro rata* with peak demand, this is also average copper loss. The second, negative, term is the sum of the reduction in copper losses in the other 8,759 hours of the year resulting from the expansion of capacity. The third term is the increase in annual iron losses, these being proportional to capacity and hence to demand in the peak hour.

Marginal annual loss with respect to an increase in load in any other hour $t \neq p$, is

$$\frac{\partial K_t}{\partial L_t} = 2\frac{C}{EL_p} L_t$$

If, now, C is taken to relate to a series of sections in a radial distribution system, e.g. from the infeed from the transmission system down to consumers' terminals (which can be done if no loads are supplied from any sections other than the last, if all sections at each level are identical and if load is equally distributed between the sections at each level) the above expressions apply to the whole of that distribution system. They can then be used to ascertain the annual energy cost of any specified load increment, given the structure of the marginal cost of generation and transmission. Thus if the load increment for $t \neq p$ is W_t kWh and the incremental generating cost adjusted for transmission losses per kWh in that particular hour is J_t, the incremental cost of supplying the load with respect to that hour is:

$$J_t \frac{W_t}{1 - \dfrac{\partial K_t}{\partial L_t}} = J_t \frac{W_t}{1 - 2\dfrac{C}{EL_p} L_t}$$

80

Investment and Cost Structure

The difficulty arises in costing an increment of demand in the peak hour, dL_p, since the resulting change in losses is spread over the whole year. Hence the net cost could only be ascertained precisely by calculating

$$-J_t \frac{C}{E} \cdot \sum_{t \neq p} L^2 \cdot \frac{1}{L_p^2}$$

for each of the other 8,759 hours.

It is evident from all this that the calculations required would be impossibly demanding in practice. On the other hand, total losses cannot be measured and related to total load in order to derive a statistical estimate of C because this would require: (i) no unmetered consumption, and (ii) synchronized reading of all meters. The latter is quite impossible. Thus the most that can be hoped for are rough estimates of incremental losses on the different voltage levels of the distribution system at different load levels.

7

THE QUALITY OF SERVICE

THE NATURE OF QUALITY

The quality of service has three ingredients: voltage fluctuation, frequency variation and reliability. They are not entirely independent, since if demand exceeds generating capacity there is a choice between reducing frequency and voltage on the one hand and cutting off supplies on the other. Bearing this in mind, let us briefly examine the causes and nature of each of the three.

Apart from voltage drop as a means of reducing load at times when there is a shortage of generating capacity, voltage fluctuations also arise as a result of changes in the load on the transmission and distribution system. A change in voltage will affect the heat output of heating appliances, can worsen television reception and the performance of X-ray apparatus and may upset the functioning of some electric motors. A rise in voltage reduces the life of electric lamps.

Frequency variations (which rarely occur apart from capacity shortage) adversely affect some industrial plant and make electric clocks inaccurate.

Interruptions of supply may be caused by available generating capacity shortage, by breakdown of transmission and distribution mains and equipment or may be necessary in order to carry out engineering work without danger, in which case they can usually be planned and announced in advance.

THE ECONOMIC PROBLEM

The nature, causes and cures of poor quality of supply constitute an enormous technical subject. From the economic point of view, however, it is sufficient to note that the quality of supply can be improved at a cost. Thus the provision of a larger margin of generating capacity in relation to forecast load, parallel circuits, larger capacity distributors, automatic reclosing devices, voltage regulators, communications facilities and so on and so forth all cost money and all act to improve the quality of supply in one way or another to some or all consumers.

The problem is, then, that a choice has to be made between selling electricity with a higher or lower quality of service at a higher or lower price. This choice has two major features: first, with a few exceptions it has to be common for all consumers; second, it is not possible to base it upon any calculus of benefits in relation to costs. To put this last point in another way, no marginal conditions formulated to describe the optimal quality level are operational.

First, consider the exceptions. A large consumer (e.g. a factory) can be provided with two separate links to the grid so that if one of them fails the other suffices to provide a continued supply. This clearly diminishes the risk of failure of that consumer's supply insofar as that risk relates to the part of the distribution network which is peculiar to him. The duplicate facility has an identifiable cost. Hence it is perfectly possible, if so desired, to let the consumer choose whether or not to have it and to charge him the corresponding capital cost.

The other exception consists of so-called interruptible or restrictable supplies. Here an electricity supply undertaking lowers its charge to a consumer in return for the right to reduce his supply at times (not fixed in advance) when demand rises close to available capacity. In practice this device is confined to some large industrial consumers.

Apart from these two main exceptions, there is no possibility of letting consumers individually choose the price–quality combination which best suits them. Thus the decision has to be made by the electricity supply undertaking. This makes it a fit subject for cost–benefit analysis. Unfortunately, this is just not a practical possibility. In order to explain why, it will suffice to confine attention to reliability, i.e. the risk of interruption, since this is the least difficult aspect of quality to analyse.

THE MARGINAL COST OF RELIABILITY

An earlier chapter explained how some measure of the future risk of load reduction due to a shortage of available generating capacity can be calculated as a function of the planned level of installed capacity. Given this functional relationship and given the addition to system costs resulting from a small increase in generating capacity, the cost at the margin of reducing the risk of load reduction can readily be determined.*

Such calculations are generally carried out for a generating system as a whole. Multi-area calculations which bring transmission capacity between areas and the risk of transmission outages into account, so that separate risk estimates are made for each area, are being developed. It appears that the calculations are too complex for exact methods so that a Monte Carlo approach is required. Thus geographically differentiated risk and cost calculations are not yet a planning tool.

The reliability of a distribution network depends on the outage rates of its components, upon the time taken to restore supply (by switching or repair) and upon whether the components are in series or in parallel. Thus in the simplest case where a number of components are in series and where component failures are random and independent, the probability of failure of supply is the sum of the probabilities of failure of each of the components. Matters get more complicated when component failures are nót independent, which is important in connection with weather-related failures (e.g. storm damage) and when there are parallel circuits. In the latter case, interruption occurs only when there is at least one failure in each of the parallel circuits or when those remaining in service are inadequate to take the load.

These complications can be handled if sufficient data are available.† This is an extremely large requirement. But even if it is fulfilled, the calculations do not relate to the distribution system as a whole but only to a network constituting a very small part of it. Thus the reliability of two alternative schemes for supplying a new housing estate might be compared in order to see whether the more

* With impeccable logic the French treat this as the *implicit* marginal cost of supply interruptions and choose to use its past value as a guide in system planning for the future. See Michel Dupoux: 'L'Ajustement des Programmes d'équipment de l'Electricité de France', *Revue Française de l'Energie*, April 1966. This is neither more nor less arbitrary than a decision to preserve a past level of reliability and its simplicity is slightly spoiled by reserving (p. 371) 'la possibilité de légères retouches à l'avenir, selon l'evolution du désir de sécurité de la clientèle'.

† For reliability calculation techniques and description of a computer programme see Montmeat, Patten and others: 'Power System Reliability', *IEEE Transactions on Power Apparatus and Systems*, July 1964 and July 1965.

costly one offers greater security. While this may be useful, it is clearly nothing like the calculation of any general marginal cost of reliability for the whole distribution system.

The circumstances in which any such general calculation can be made appear to be limited. Where the product of failure rate of a particular type of component and the repair or switching time required to deal with its failure are both known and enter additively into the expression for total hours lost, the gain in reliability which would be furnished by a reduction in its failure rate or repair or switching time can be calculated without the need for other data. But there is no presumption that other system improvements would not furnish an equal gain at less cost.

These remarks make it clear that the allocation of expenditure between generation and transmission and distribution so as to maximize the reliability achieved for a given total system cost is not yet an operational notion. This would be true even if there were some single-valued measure of reliability which reflected its value to consumers.

THE VALUE OF RELIABILITY

The frequency of supply interruption, hours lost and kWh lost—all on a per consumer per year basis—are all single-valued measures of reliability. None of them, however, is obviously proportionate to the social cost of interruptions as the following conundrums make clear:

— Which is worse, losing 100 similar consumers for one hour or one such consumer for 100 hours?
— Which costs more, interrupting 100 MW of industrial load or 100 MW of domestic load?

Attempts to infer the valuation of reliability from the expenditures which consumers deem it worth incurring to provide standby supplies are impossible to undertake, because these consumers' subjective estimates of the reduction in risk thereby secured are unknown. In any case, such calculation would yield no information about the vast majority of consumers.

CONCLUSION

Determination of the security of supply must rest on judgement rather than upon calculation. Costs will reflect the standards of security chosen.

8

MARGINAL COSTS AND OPTIMAL PRICING

THE PROBLEM

Marginal cost pricing in electricity means a tariff structure such that the cost to any consumer of changing the level or pattern of his consumption equals the cost to the electricity supply industry of his doing so. This can be achieved more or less closely according to whether the tariff structure is more or less complicated.

This chapter examines in turn the case for marginal cost pricing, the complications introduced by financial constraints and the question of short-run versus long-run marginal costs. It concludes that there is a case, that financial constraints may provide financial discipline for management at the expense of efficient resource allocation and that short- and long-run marginal cost pricing are equivalent. The issue of whether the tariff structure should be simple or complex is postponed until the next chapter.

OPTIMAL PRICING

Marginal cost pricing, as defined above, will cause the interest and hence the electricity consumption decisions of consumers to conform to the national interest if:

(*a*) consumers are sensible and well-informed
(*b*) the distribution of wealth and income is acceptable
(*c*) cost to the electricity supply industry of meeting changes in consumption coincides with the value to the economy of the resources used or freed, i.e. social opportunity cost
(*d*) the prices of substitutes and complements for electricity (in both final and intermediate consumption) equal their marginal social costs as do the prices of goods and services which use electricity in production.

Now many readers will regard these conditions as not being met in whichever economy happens to interest them. Nonetheless, I believe that there is a presumption in favour of marginal cost pricing in most economies with a fairly freely functioning price mechanism.

Consider first condition (a). On the whole, people and institutions are neither grossly ignorant nor swayed by prejudice in purchasing energy. In any case, to the extent that they are ill-informed, the electricity industry (and other energy industries) can cure this, so far as their own sales are concerned, by informative rather than persuasive advertising, by good technical advisory services and so on.

If condition (b) is not fulfilled my view is that the electricity industry should not attempt to put matters right. It may be told, by government, to subsidise one or other group of consumers but otherwise should act *as if* the existing distribution of income and wealth were acceptable.

If these judgements are accepted, then the relationship to be sought between the structure of electricity tariffs and the structure of electricity marginal costs depends upon the relationship between prices and marginal costs elsewhere in the economy and upon external economies and diseconomies. Only if both all other prices are equal to the respective marginal costs and all external economies and diseconomies have been taken care of will marginal cost pricing for electricity be *a priori* in the national interest. But these twin conditions are not fulfilled in practice, so the problem is one of sub-optimization—second best.

In order to tackle this problem in practical terms I suggest that since it is impossible to take account of what is unknown and difficult to take account of many minor non-optimalities, the right policy is to pursue marginal cost pricing for electricity subject to 'corrections' made only for those non-optimalities which are known to have a significant effect on the demand or cost structure of electricity.*

* In an interesting application of modern theorizing on second-best problems M. J. Farrell has recently argued that since there are more indirect taxes than subsidies and since few private businesses sell at less than marginal cost there is a *general* presumption that the optimal prices to be charged by public utilities are *not less than* their marginal costs except where price/marginal-cost divergences are heavily concentrated on close complements. As he recognizes, however, while this applies to gas and electricity jointly, it may not apply just to one of them taken separately. Furthermore, electrical appliances, which are certainly close complements, are heavily taxed in some countries and incentives (subsidies) to private capital formation frequently cover the installation of private generating plant which provides a nearly perfect substitute for public supply. Farrell's paper 'In Defence of Public-Utility Price Theory' is to be found in *Economics of Public Enterprise*, Penguin, 1968.

The sort of non-optimality which has to be disregarded can be illustrated by two extreme examples. Take first, the existence of a large excess of the price of shaving cream over its marginal cost. This will make the use of electric razors larger than it would be in a 'first-best' situation, but it would nevertheless be absurd to suggest adjusting the price of electricity to compensate for it. The non-optimality is, by assumption, known but is totally insignificant. Second, consider the point that income taxation may cause labour inputs to be non-optimal. Since this affects the whole economy the effect is not trivial, but as its implications for electricity are wholly unknowable there is no point in fretting about it.

The factors which will justify pricing electricity at above or below marginal cost can now be listed as follows. Expressing the matter in terms of the reason for an excess, they are:

(*a*) important close substitutes sell at significantly above marginal cost or generate large external economies
(*b*) products in whose production electricity constitutes a major input sell at significantly below marginal cost or involve large external diseconomies
(*c*) important close complements sell at significantly below marginal cost or generate large external diseconomies
(*d*) major inputs of the electricity industry are bought at significantly below marginal cost or involve large external diseconomies.

It is immediately obvious that while it will be desirable to adjust electricity pricing to allow for these non-optimalities it would, in many cases, be better still to tackle them directly. The choice in any particular case will depend upon who is in a position to do what. This brings in all sorts of considerations.

Consider, as one possible case, a situation where there is no winter–summer differential in gas prices despite a large excess of winter marginal cost over summer marginal cost. If this non-optimality is entirely and permanently unalterable it constitutes a reason, *ceteris paribus*, for introducing a similar divergence in electricity pricing for sales in competing uses. But if there is a chance that it could be altered, those responsible for electricity pricing have to allow for the possibility that if they allow for the non-optimality they are likely to perpetuate it, while if they do not they may hasten its removal. The likelihood of this will, of course, vary from case to case, so generalization is impossible. My personal bias is towards aiming for the best rather than adjusting to the second best. This has the subsidiary merit of simplicity since it requires less knowledge on the part of those responsible for electricity pricing. Thus, to revert to the particular example just given, an electricity industry cannot be expected to know much about the economics of the gas industry.*

Where non-optimalities of the sort listed above are unalterable, however, they justify departures from marginal cost pricing in electricity. Since such departures depend on the particular circumstances of particular cases they have to be left out of any general discussion and the next chapter, which discusses tariffs, accordingly assumes that full-blooded marginal cost pricing is aimed at. Even where circumstances are such as to make this inappropriate, the following discussion is still necessary as a starting point in formulating tariffs, since the desirable departures from marginal cost pricing will be expressed as mark-ups or mark-downs on marginal costs.

FINANCIAL CONSTRAINTS

Marginal cost pricing, subject to any appropriate adjustments for non-optimalities, may yield a revenue which provides a surplus in relation to accounting cost which is too high (in the case of a regulated utility subject to a ceiling on earnings) or too low (in the case of a public enterprise which has been set a target minimum of earnings). If this financial constraint is sacrosanct then some prices will have to be set below marginal costs (in the first case) or above (in the second). In either case a constraint which is effective will result in a welfare loss.

This does not mean, however, that such a constraint is necessarily undesirable. Thus a constraint of the second kind, a financial target to be aimed at or exceeded, can serve to provide some check upon the management of the electricity authority and to make it cost-conscious. It may be considered that the resulting gain in managerial efficiency outweighs the welfare loss just mentioned.

Given the existence of such a constraint, the problem arises of minimizing the welfare loss involved in meeting it. This problem is

---

* If some third party such as a government department has to provide the necessary information and urge the electricity industry to take account of it, there seems no reason why that third party should not seek to secure a change in gas tariffs rather than electricity tariffs.

a complicated one when there are demand and cost interdependencies between different times in the tariff structure, when the electricity industry competes with, buys from and sells to other public enterprises (such as coal) which have their own financial constraints and, finally, when there are price/marginal-cost divergences in the private sector. Mr Ray Rees has provided a general theoretical analysis which takes into account all these complications in his paper 'Second-Best Rules for Public Enterprise Pricing'.* It emerges quite clearly that the solution which he provides requires far more knowledge of elasticities and cross-elasticities than will ever be available. Once again, therefore, as Mr Rees points out, all that is practical is to take account of the factors that are both known and important. In the case of least knowledge this will boil down to an injunction to set the proportionate excess of price over marginal cost required by the constraint on different items according to 'what the traffic will bear' except in the case of sales by one financially constrained public enterprise to another.†

The optimist will endeavour to persuade those responsible for imposing a financial constraint to choose its form and level so as to provide the necessary incentives and checks for management without requiring any substantial departure from an optimal pricing

* *Economica*, 1968.

† Mr Rees's formal analysis shows it *not* to be true in general principle that all such sales should take place at marginal cost even when privately produced substitutes and complements are available to the purchasing public enterprise at marginal cost. This is a surprising result because it means the pricing rules which secure constrained overall welfare maximization by the two public enterprises do not necessarily involve the purchasing enterprise in minimizing the social cost of whatever constellation of outputs it ends up by producing. The reason is that the welfare gain to be had by its moving to social cost minimization could conceivably be more than offset by other adverse repercussions on welfare of three sorts. First, the change of revenue which the selling public enterprise would sustain by altering the price on its sales to the purchasing public enterprise to equality with marginal cost would have to be compensated by changing the margin between prices and marginal costs on its other sales, with consequent welfare effects. Second, the change in the purchasing enterprise's inputs of private sector products could react back on the demand for either enterprise's products (or supply of factors) with consequent welfare effects when they are not being sold at marginal cost. Third, a change in either of the public enterprises' inputs or outputs which alters the equilibrium output of a privately produced good may involve a welfare loss even though the latter is priced equal to marginal cost because this equality may not be optimal when related public enterprise products are not priced at marginal cost.

When the principles of ignoring the trivial and the unknowable cause these three classes of effects to be left out of account, sales by one public enterprise to another should all be at marginal cost.

structure. Some departure is inevitable if the constraint is effective, but it need not be large.

LONG-RUN AND SHORT-RUN MARGINAL COSTS

If the case for marginal cost pricing is accepted, the question remains whether it is long-run marginal cost (as defined in this book) or short-run marginal cost that is relevant. This question, which was the subject of some controversy in post-war theoretical discussions of the pricing policy of public enterprises, turns out to be an unreal one. Short- and long-run marginal cost pricing are in fact equivalent given correct forecasting, i.e. granted that the assumptions made about the future in calculating long-run marginal cost turn out to be valid.

This equivalence, established by Boiteux and other French writers on public enterprise economics,* requires a particular interpretation of short-run marginal cost pricing and also requires supplementation by an investment principle. The particular interpretation is that when demand exceeds capacity, price is set at the level necessary to restrict demand to capacity output. Thus the short-run pricing rule is as follows:

> The price of electricity at each time, place and voltage is set equal to marginal running cost per kWh delivered *or*, if it be higher, at the level necessary to restrict demand to capacity.

The justification of this interpretation is simply the judgement, which will be taken for granted here, that rationing by price is preferred to rationing by power cuts.

Since this pricing rule relates only to the short-run, taking capacity as given, it needs to be supplemented by a long-run rule which determines the optimal capacity. A general formulation of this rule, the investment criterion, which has the same sort of justification in terms of optimal resource allocation as the short-run pricing rule, is as follows:

* Some of their papers are to be found in translation in *Marginal Cost Pricing in Practice*, ed. J. R. Nelson, Prentice-Hall, 1964. See also Jacques Drèze: 'Some Postwar Contributions of French Economists to Theory and Public Policy', *American Economic Review*, Part 2, June 1964. Another statement of the equivalence between short- and long-run marginal cost pricing is given by Herbert Mohring on pp. 233–239 of his contribution to *Measuring Benefits of Government Investments*, ed. Dorfman, Brookings, 1965.

An investment is undertaken if the present worth to consumers of the consequential change in their supplies, less the present worth of the difference in total system costs with and without the investment is positive.

Where an investment involves no change in supplies, this simply reduces to the statement that investments which lower the present worth of system costs are worth undertaking.

An investment will cause a change in supplies (as compared with what they would have been in its absence) if it lowers off-peak marginal running costs or expands capacity, since either of these things, by the operation of the short-run pricing rule, will lead to a reduction in prices. The effect of price changes upon consumption is complicated; it seems that in most cases prices affect appliance ownership rather than the use made of a given stock of appliances. Thus distributed lags are involved and we cannot regard consumption in any half hour as primarily determined by the price in that particular half hour. We do not need to go into these matters here, however. We are assuming that prices are fixed according to the short-run pricing rule and are concerned only with measuring the value of changes in consumption induced by price changes consequent upon its operation. Any change in consumption can be valued at what consumers pay for the marginal unit. With an increase or decrease in this of dp from p to $(p+dp)$ in any half hour and a change in consumption during that half hour from U to $(U+dU)$ the approximate value to consumers of the change in supplies is $dU(p+[dp/2])$. Any change in consumption during half hours when price has not changed must be valued at what consumers pay for the marginal unit in those other periods, $p \cdot dU$. This is all rather approximate, as it assumes linearity, but sophisticated analyses in terms of the variants of consumer's surplus are too non-operational to be of interest.

DEMONSTRATION OF THE EQUIVALENCE

The long- and short-run pricing rules both require the consumer who takes a specified load increment to pay the incremental running costs involved in providing the extra kWh, namely $PW[dUm]$. In addition:

(a) the long-run rule requires him to pay dP times the present worth per kW of incremental capacity costs;

(*b*) the short-run rule requires him to pay d*P* times the excess of price over incremental running cost per kWh which is necessary to restrict demand to equal capacity at times of peak.

Now the revenue obtained (in excess of marginal running cost) per kW of capacity under (*b*) measures the value to consumers (in excess of marginal running cost) of the peak kWh which they would lose if capacity were 1 kW less. The investment criterion requires, among other things, that the present worth of this change in their supplies just equal the present worth of the difference in system costs with and without that marginal kW of capacity. This, however, is the present worth per kW of incremental capacity costs paid under (*a*). Thus (*a*) and (*b*) are equivalent, in other words the investment makes the quasi-rents under (*b*) equal to what is required under (*a*).

This equivalence, as stated above, rests upon the assumption that anticipations are correct. Since the time period involved in planning and installing new capacity is some five or six years, the assumption will rarely be met. Thus if demand turns out to have been underestimated, so that in the event capacity is less than required, the short-run pricing rule will require higher peak prices than long-run marginal cost. But this is just one of the risks which require the planned level of capacity to exceed the mean forecast level of peak demand. Where a margin of capacity is planned to provide against these risks, it should be possible to meet the unforeseen demand without raising the price of peak power. Thus the above argument still holds so long as it is couched in terms of mean expectations.

It may be argued that if, in the event, the plant margin turns out to be larger than is necessary to provide the chosen level of security against weather and availability risks, then prices should be lowered so that the excess plant margin is used up, and conversely. Whether this is possible depends partly upon the way in which tariffs are fixed. But whatever the possibilities are, the main point is unaffected. Long-run marginal cost pricing is, in terms of mean expectation, equivalent to the combination of short-run marginal cost pricing and the optimal investment rule.

9

TARIFFS

THE RANGE OF POSSIBLE TARIFF TYPES

The electrical quantities relating to the load of a consumer which are measured in a tariff context are only four in number, namely voltage level, demand (i.e. kW or kVA), units (i.e. kWh) and Power Factor. Nonetheless the number of different possible tariffs that can be applied to any particular class of consumers is enormous as the following classification shows. It is not exhaustive but does serve to indicate the range of possibilities.

First, tariffs may be related to such conditions of supply as voltage level and position in the system either geographically (e.g. rural versus urban) or electrically (e.g. whether or not the consumer provides transformers).

Second, the supply may be restricted or unrestricted. Restriction may be:

(*a*) by time:
    (i) fixed in advance (e.g. night only), or
    (ii) at short notice (i.e. interruptible supplies), and/or
(*b*) by use (e.g. the old separation of domestic heating and lighting).

The restricted-hour domestic tariffs in the UK often involve both forms of restriction, e.g. when confined to storage uses.

Third, there may be a demand (kW or kVA) rate. If there is it can be related to:

(*a*) a contracted maximum demand (as in the 'fuse tariff' where a circuit breaker interrupts the supply if the consumer exceeds his contracted demand), or
(*b*) the measured maximum demand of the consumer, annual, monthly, etc., or
(*c*) the measured demand of the consumer at certain times specified in advance (e.g. potential peak hours).

94

The demand charge per kW or kVA may be constant or it may be related to the size of that demand or to some other factor such as the ratio of average to maximum demand in the year. It may vary seasonally or monthly, etc. It may also be related to Power Factor or supplemented by an additional charge so related.

Fourth, in most cases there is a unit (kWh) rate, but this is not universal. This rate may:

(*a*) not be timed, or

(*b*) vary seasonally or monthly and/or by time of day and week.

In either case the rate may be unique or it may vary with the number of units (as in a block tariff), with installed load or demand as in the previous item, or with some other factor such as the size of the consumer's establishment or an assessment of his maximum power requirement. It may also be varied to reflect changes in the prices of fuel or other inputs.

Finally, there may or may not be fixed or variable charges related to some non-electrical quantity (e.g. floor area).

VOLTAGE LEVEL AND LOCATION

As a first step in enunciating principles for making choices within this vast range of possible tariffs, cost differences related to voltage level and to geography may be considered.

In general, it is cheaper to meet a given magnitude of load at a given place at high voltage than at low voltage. There is no need for transforming points to step down to low voltage and high voltage mains have a greater carrying capacity than low voltage mains. Capital costs are therefore lower and, in addition, distribution losses are smaller. Hence whether or not the structure of high voltage tariffs is similar to that of low voltage tariffs, it is appropriate that their general level should be lower for consumers of a given size in a particular location. In practice, the overlap is limited, since only industrial consumers and some large commercial consumers take high voltage supplies and since scarcely any of the largest industrial consumers take low voltage supplies.

In an integrated system, geographical cost differences at any given voltage level spring from three sources, as is apparent from earlier chapters. First, there may be differences between nodes in the incremental cost of peak capacity if (but only if)* the cost of all new

* See pp. 48–49.

plant varies geographically or the peaks at different nodes do not coincide. Such differences can arise, in other words, only when there are locational constraints on new plant or when there is diversity. Their magnitude will be larger the greater is this diversity and the greater the difference in generating cost conditions between the nodes. In general, therefore, they will be larger the farther apart are the nodes. Thus in an extreme case, these differences between incremental peak costs might be important between a cold north with hydro power and a warm coastal south with cheap oil. At the other extreme, they would be negligible in a country where new nuclear and gas turbine plants can equally well be located in any region and where regional differences in the mix of different types of load are small.

Geographical cost differences in an integrated system may, secondly, consist of differences in incremental running costs; these cannot be more than a few per cent in the absence of constraints upon transmission.

Third, distribution costs, as explained in an earlier chapter, are inversely related to load density. This means that they differ between urban and rural areas rather than between regions.

The reflection of geographical cost differences in tariffs is subject to an extremely important practical limitation. This is the problem of drawing boundary lines between cost zones. On the one hand, neighbouring nodes have to be grouped together, both in order to avoid complexity and because the layout of the distribution network (i.e. which consumers are served by which node) is constantly changing as this network is expanded and as new nodes are introduced. On the other hand, the larger are the groupings which are treated as units for cost purposes, the greater will be the cost differences between adjacent groupings, with arbitrary tariff differences resulting for consumers on either side of the boundary line. Thus whether or not regional tariff differences are desirable, they are only practicable either if physical features provide a natural boundary line or if pre-existing and firmly established administrative boundary lines exist. In a two-tier system, for example, where a number of distribution undertakings are supplied by a common generation undertaking, each distribution undertaking can easily have its own separate tariffs.

The practical problem of distinguishing urban and rural areas within a region is even greater, though for similar reasons. Their boundaries are far from clear geographically and are always changing electrically as the distribution system is extended and reinforced. Any distinction between urban and rural areas for tariff purposes will thus be arbitrary and seem so to consumers and will require revision every few years.

The benefit from avoiding these difficulties has to be set against the cost. The nature of this cost is shown in the accompanying diagram on the simplifying assumptions (*a*) that the system load factor is the same for all consumers so that kWh and peak kW are in a fixed proportion, (*b*) that costs are constant, and (*c*) that demand

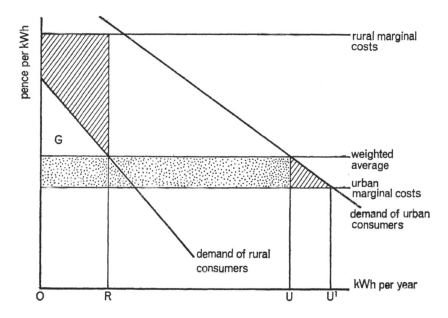

is not growing. If both classes of consumers are charged a weighted average of urban and rural marginal costs, urban and rural consumers would consume OU and OR kWh per annum respectively.* If each had been charged their own marginal cost, rural consumption would have been zero and urban consumption would have been OU^1 kWh per annum. This would have raised the sum of consumer and producer surpluses by the amount of the two cross-shaded areas:

— the area on the right representing the excess of the value to urban consumers over the extra cost of the increase UU^1

* By construction, the dotted area equals OR times the excess of rural over weighted average marginal cost.

— the area on the left representing the excess of the cost saved by not supplying rural consumers over the value to them of their consumption *OR*.

If, on political grounds, rural electrification and a uniform tariff are chosen, the implicit judgement is that it is worth sacrificing these gains:

(a) in order to transfer the dotted rectangle of urban consumers' surplus to rural consumers
(b) to avoid the arbitrariness of establishing an urban–rural boundary line.

Another way of putting this is to say that to avoid arbitrariness and to give rural consumers a net gain of the triangle *G* is politically preferable to giving urban consumers a net gain equal to the dotted rectangle plus the shaded triangle on the right.

In the rest of this chapter it is assumed that the choice has been made and the analysis relates to an area or region within which tariffs are uniform for each class of consumer.

THE STRUCTURE OF A COST-REFLECTING TARIFF

If the case for marginal cost pricing be accepted, if the cost and bother of charging consumers were unaffected by tariff complexity and if there were no difficulties in altering tariff structures, then the right tariff structure to adopt follows from the analysis of costs in previous chapters. It is useful to set this out to serve as a reference point before taking account of the cost and bother which such a structure would involve. It would contain the following components:

(a) a connection charge for each new consumer equal to the capital cost wholly and exclusively incurred in connecting his premises to the system
(b) a rate per kW of contribution to peak at each voltage level of the distribution system, equal to the average annuitized cost per kW increment of capacity at that voltage level grossed up by the capacity margin.*
(c) an annual charge equal to the marginal annual cost per consumer of consumer service, meter-reading, billing and account collecting
(d) a contribution to general overheads and to the annual cost of running, repairing and maintaining the system

* These charges might vary with Power Factor or might be levied per kVA.

(e) marginal running cost per kWh, grossed up to allow for incremental transmission and distribution losses, and varying from hour to hour as the total generating load changed

(f) incremental capacity costs per kW of contribution to peak generation, calculated according to the concepts spelled out on pp. 54 ff.

It will be noted that this formulation allows for different costs at different voltage levels. Thus the structure as set out could be applied to all consumers without any need to distinguish between them on account of non-electrical characteristics. One and the same complex tariff structure would serve for everyone from the owner of a week-end cottage to a steel mill. This proposition is of interest because it makes it clear that the need to distinguish consumer classes and the criteria for doing so stem from the need to avoid complexity in tariffs.

WEATHER SENSITIVITY AND PEAK PRICING

The fact that the load is weather sensitive means that the level of the load in any hour involves a random element. Thus marginal running costs and the level and timing of distribution and generation peaks also involve a random element.

It is not practicable to raise or lower the kWh rate at short notice when extra cold or mild weather raises or lowers the load from its mean level for the time of day and year and so raises or lowers marginal running costs. Even if it were practicable there would be no point in it unless consumers could be immediately informed and so induced to react. Thus item (e) above, marginal running cost, has to be estimated in advance so that kWh rates can be fixed in advance to reflect the regular seasonal, weekly and daily load curves. These rates and the times to which they relate have to be calculated for average weather conditions, consumers thus shifting the risk of weather variations around the normal seasonal pattern on to the electricity industry. This risk, as explained earlier, affects the plant margin necessary to provide any given level of security and is therefore reflected in capacity costs.

The fact that, because it is affected by weather, the timing of the annual peak can be forecast only a few hours in advance has important consequences. The aim of charging items (b) and (f), the capacity charges, is to provide consumers with an appropriate incentive to economize in their use of power at peak and there is no point in providing such an incentive to consumers who are unable to respond to it. If follows that a useful purpose is served by levying the kW rate on a consumer's contribution to the actual peak only when, first, the consumer can be given a warning that a peak is expected and, second, the nature of his use of electricity is such that he can readily restrict his demand at short notice. Now the cost of installing and operating a warning system is such that it is only justified when the amount by which the consumer will restrict his demand is fairly large. It follows that capacity charges levied on the contribution to actual peak can be confined to large consumers (industrial firms) whose processes are interruptible without incurring a big increase in production costs. Users of certain electric furnaces provide an example.

For all other consumers, i.e. for the vast majority, it is not the actual peak which matters but potential peak hours, the set of hours within which the actual peak is expected to fall. In probabilistic terms each such hour has a certain probability of turning out to be the actual peak, the probabilities summing to unity.

In order to proceed by means of a simple arithmetic example, let it be supposed that:

— there are 200 potential peak hours, each with an equal probability of being the actual peak
— the appropriate rate per kW of mean expected demand during these 200 hours is £10 = 200s. (This is the mean expected demand in these hours in average weather conditions and is consequently easier to forecast than actual peak demand.)
— among a group of similar consumers the sum of their individual maximum demands, on past experience, is 1.2 times the maximum demand of the group as a whole during the 200 hours
— on average the actual peak is expected to be 1.07 times the mean expected demand during the 200 hours.

Then the following three ways of recovering the peak capacity charge from this group of consumers are all equivalent, both in providing the same expectation of revenue and in giving these consumers an identical incentive to economize in their use of peak kW:

(a) £10/1·07 per kW at time of actual peak

(b) £10/1·2 per kW of each consumer's own maximum demand during the 200 hours

(c) 1s. per kWh surcharge on the running rate for all units taken during the 200 hours

(a) as argued above, is required for those consumers who can be warned when the actual peak is expected. For all the other consumers (b) and (c) will do just as well. Since the cost and bother of charging along these lines is less than that of charging according to (a), either (b) or (c) is preferable.

Although the example is simple, it brings out an important point. In alternative (b) the appropriate charge depends upon the demand diversity among the group of consumers. If, in another group, the sum of their individual maximum demands were greater in relation to the maximum demand of the group as a whole, say 1·35, then the appropriate charge would be lower, namely £10/1·35. Thus if (b) is chosen because the cost and bother is less than in the case of (a) or (c), there may be a need to distinguish different groups of consumers, each group paying a different maximum demand charge according to its demand diversity.

What has just been said implies that, however the peak capacity charge is recovered, the amount should be uniform over all the potential peak hours. This simplification made it easy to explain the significance of the random element in the peak.

More generally, however, the potential peak period is defined as including all those hours when pricing at marginal running cost alone (i.e. without the addition of a peak capacity charge) would lead to an excess of demand over capacity. This means that under average weather conditions for the time of year and apart from random fluctuations, demand would exceed installed capacity minus the chosen capacity margin. The function of the peak capacity charge is to prevent this excess occurring, and it must be higher the greater would this excess otherwise be. Its level may thus ideally vary from one group of potential peak hours to another. If, for instance, the evening peak would exceed the morning peak were they both charged the same, the capacity charge must be higher in evening potential peak hours than in morning potential peak hours.*

---

* For a recent restatement of peak pricing theory see Oliver Williamson: 'Peak-load Pricing and Optimal Capacity under Indivisibility Constraints', *American Economic Review*, September 1966, reprinted in *Economics of Public Enterprise*, Penguin, 1968.

To ascertain the ideal time pattern of peak capacity charge would thus require very considerable knowledge of consumers' reactions to a change in tariff structure. Such knowledge is in fact very limited. It is therefore fortunate that these consumer reactions are very slow, so that peak-shifting does not in practice present an intractable problem.

In practice, indeed, the difficulties of changing existing tariffs, the continual evolution in the markets for energy, technical developments affecting the cost structure and changes in the general price level mean that what can be done and what needs to be done are usually changing. This fact, coupled with the difficulty of foreseeing consumer reactions, means that the principles analysed in this chapter will never be capable of precise application. Their role is to provide a standard of reference and to inform judgement, not to map out precise proposals.

SIMPLIFICATION

A tariff structure which fully reflected the structure of marginal cost and included an ideal time pattern of the peak capacity charge would be extremely complex. This has been made clear in the last two sections above. Complexity, however, is both expensive and inconvenient. Thus even if knowledge of future developments and consumer reactions were complete, it still would not be sensible to introduce as complex a tariff structure as that described above.

The point is, then, that the cost and bother, both to the supplier and to the consumer, of metering, billing and accounting depend upon the complexity of the tariff structure. These costs can be reduced by simplifying tariffs at the expense of giving consumers a less detailed message designed to influence their electrical behaviour. How much this matters depends upon the extent to which consumers' electrical behaviour is influenced by tariff structure. The problem is to balance savings in cost and bother against the loss from departing from the ideal tariff structure.

At one extreme, consider the supply of electricity to a telephone kiosk using a 40 W bulb. Even if the Post Office would be induced to switch it off during potential peak hours by the imposition of a peak kW charge, the resulting saving in generating capacity costs would clearly be far less than the cost of the necessary metering. Indeed even an ordinary kWh meter would cost too much to make a kWh rate preferable to an annual fixed charge based on estimated consumption.

At the other extreme, consider a large industrial consumer with a load measured in tens of megawatts and able to restrict demand at short notice or at predetermined times in order to save money. The cost of levying even an extremely elaborate tariff is clearly trivial in relation to its advantages.

This intuitive example shows that the size of consumer response to tariff structure and the cost of metering, billing and accounting are relevant to choosing the appropriate degree of simplification. What is needed, then, is to show how these two factors can be brought into an analysis of marginal cost pricing. The key lies in the fact that marginal cost pricing involves maximizing the sum of consumer and producer surpluses.

In order to show this, consider a simplified arithmetic example, this time relating to domestic consumers. Assume that the weighted average marginal cost of supplying electricity to domestic consumers is 0·5 pence per unit, i.e., per kWh, at night, 1·75 pence per unit during the day and 1·50 pence per unit for the twenty-four hours regarded as a whole. Suppose that only two alternative tariffs are in question; either a flat-rate domestic tariff of 1·50 pence per unit or a two-rate time of day tariff with night and day rates of 0·5 and 1·75 pence per unit respectively. The latter requires a two-dial meter and a time-switch, so costs more to administer than the flat-rate tariff which only requires a single dial meter.

Maximization of the sum of consumers' and producers' surpluses means that any increase (decrease) in consumption is desirable if the value to the consumer of the increase (decrease) exceeds the extra cost (falls short of the cost saving) involved in supplying him. Thus consider shifting a single consumer from the flat-rate to the time of day tariff. Assuming linearity in his demand function:

— the value to him of the consequent annual change, D, in his day consumption is $\left(\dfrac{1·75+1·50}{2}\right)D$; the resulting change in the annual cost to the industry is $1·75D$. Adding these, the net annual gain is $D\left(\dfrac{1·75+1·50}{2}-1·75\right)=-0·125\,D$. D may, of course, be negative

— similarly, the value to him of the annual change, N, in his night consumption is $\left(\dfrac{1\cdot50+0\cdot50}{2}\right)N$; the resulting annual change in the cost to the industry is $0\cdot50\ N$.

The net annual gain is therefore $N\left(\dfrac{1\cdot50+0\cdot50}{2}-0\cdot50\right)=0\cdot50\ N.$

The total net effect is thus:

$$0\cdot50\ N-0\cdot125\ D-K$$

where K is the extra annual cost, in pence, of billing and metering under the time of day tariff. If this net effect is positive then this more complicated tariff is to be preferred.*

Clearly the choice cannot be made unless it is known how the consumer reacts to the change in tariff structure. Thus a knowledge of night and day demand elasticities with respect to both night and day prices is required even when, as assumed here, marginal costs are constant. The difficulties of this need no emphasis.

Let us suppose that hunch or experiment tells us that the simpler tariff is the better one. The problem remains of estimating the ratio of night to day consumption in order to get the right weighted average (here assumed to be $1\cdot50$) of $1\cdot75$ and $0\cdot50$ pence per unit. In order to illustrate this sort of problem, consider a more complicated example.

Suppose that the marginal costs of supplying industrial high voltage consumers in a given area are as follows:

£8 per kW taken at time of system peak
$1\cdot2$ pence per unit consumed in day hours
$0\cdot4$ pence per unit consumed in night hours

(Once again, the figures are made up and various complications have been left out.) Assume that only simple kWh and Maximum Demand†

---

* Analogous analyses applying much the same criterion are to be found in *Road Pricing: The Economic and Technical Possibilities*, Ministry of Transport, H.M.S.O., 1964, and in J. J. Warford: 'Water Requirements: The Investment Decision in the Water Supply Industry', *The Manchester School*, January 1966. A simple diagrammatic treatment of the argument above is not given since it requires independence of the day and night demand curves. Note that this rough and ready approach requires modification to the extent that other goods and services which are quantitatively important substitutes or complements exhibit significant divergences of prices from marginal costs.

† In electricity supply, 'demand' means the value at a given moment of the power absorbed by the consumer. It is expressed in kW or multiples thereof.

meters are to be used and that they are to be read monthly. How then can a tariff with only monthly kWh and monthly maximum demand components best reflect the structure of marginal costs as given above?

The time of system peak is not known in advance. Suppose that it is equally likely to occur in each of the four winter months, i.e. that the probability of its occurring in any one of them is 0.25. Suppose, further, that after a great deal of effort in studying the load curves of a representative sample of the consumers it has been discovered that the ratio of their aggregate demand at time of system peak to the sum of their own maximum demands is 0.75. Then a maximum demand charge imposed for the four winter months only, at a rate of £8 × 0·25 × 0·75 = £1·5 per kW of maximum demand in each month would approximate marginal capacity costs.

Turn now to the unit (kWh) charges. If, on average, the consumers in the sample took 70% of their units in day hours, the appropriate unit rate in the tariff would be 0·7 × 1·2 + 0·3 × 0·4 = 0·96 pence per unit.

If the statistics gathered from the sample are at all adequate they will probably show that this solution is too simple in two respects. First, the above relationships may vary from month to month. Second, and more important, both

— the ratio of consumers' demands at system peak to their own maximum demands (averaging 0·75), and
— the ratio of night units to total consumption (averaging 0·30).

will be an increasing function of load factor, i.e. of monthly unit consumption per kW of maximum demand. In other words, those consumers whose monthly consumption is above average in relation to their own maximum demands will be those with a high ratio of demand at system peak to their own maximum demand and with a high proportion of night units. Hence a better reflection of the structure of marginal costs will be obtained by making both

— the monthly kW charge, and
— the unit rate

increasing functions of monthly unit consumption per kW of the consumer's own maximum demand in the month.

Suppose this is done. Then there are sure to be some consumers whose load curves are such that the approximation to marginal costs furnished by the tariff is particularly poor. The question then arises whether a different tariff involving more complicated metering should be introduced for such consumers.

Although these examples leave out many of the factors involved in tariff-making, they suffice to illustrate certain points of principle. These, to sum up, are as follows:

1. Complexity in the tariff and the number of consumer classes (i.e. different tariffs) are partial substitutes. Thus in the extreme case of a tariff which is sufficiently complex to reflect the cost structure in full detail, no consumer classes are necessary. At the other extreme of tariffs consisting only of an annual fixed charge and a unit rate, the latter must be a weighted average of marginal cost at different hours and the weights will differ from one kind of consumer to another.
2. The optimal degree of complexity (and/or number) of tariffs can be derived from the aim of maximizing the sum of consumers' and producer surpluses, given sufficient information.
3. The amount of information required to calculate the appropriate level and structure of a tariff is not necessarily less with a simple tariff than with a more complex one.

Finally, it is worth noting a quite different reason for distinguishing consumer classes. So far the analysis has assumed that marginal cost pricing is the aim and has dealt with the issue of whether this should be complex or rough and ready. If the aim is to meet a financial constraint which involves some excess of price over marginal cost other issues arise. To generate the excess revenue in a way which is considered fair and which is administratively simple but which involves the minimum departure from the resource allocation which marginal cost pricing would secure may require different kinds of consumers to be treated differently. For example:

(i) arguments about fairness are different as regards households and businesses
(ii) some indices of ability to pay, such as number of rooms, are appropriate or feasible for some types of consumer but not for others
(iii) a high standing charge or first-block rate may choke off some types of consumer but not others.

Tariffs

SUMMARY

It would be legitimate to proceed from the last chapter's case for marginal cost pricing and the analysis on pp. 98–9 of the structure of marginal costs combined with the theory of peak load pricing to propositions about optimal tariffs only if differences in the cost and bother of charging different tariffs were ignored. Any practicable tariff applicable to large numbers of small consumers can only reflect costs approximately, averaging between different consumers and over different periods of the year. The consequent necessity for choice between alternative simple tariffs is not itself simple. Even if change itself involved no costs and no distributional questions were involved, tariff-making aimed to secure optimal resource allocation would still be rendered imperfect by a lack of information.

NOTE: Some further material on tariffs is contained in the author's paper 'Peak Load Pricing', *Journal of Political Economy*, February 1968.

THE ECONOMIC ANALYSIS OF CHOICE
OF TRANSFORMER

This paper first enunciates some general principles of investment choice; it then describes the costs involved in using a transformer and, finally, explains how some particular problems relating to choice of transformer at a given point within a given system can be solved. This is a matter of sub-optimization of transformers only, finding the cheapest way of transforming at a single-transformer substation in order to supply a demand having a specified load curve. In other words the problem is when to install which transformer for how long, with a given forecast of load growth. The principles suggested also apply to the slightly broader problem where substation site and building costs are also included. They are useful for working out general rules rather than for repeated application to individual cases. The analysis makes no claim to being original, but it is hoped that a systematic statement will be useful. A number of colleagues in the electricity supply industry have helped me in writing it.

PRINCIPLES OF ECONOMIC APPRAISAL

Present Value

Given a load forecast, the engineer's task is to minimize the present value of system costs involved in supplying that load, subject to meeting given security requirements and subject to such unquantifiable practical factors as may be relevant. The costs of doing anything are measured as what would be saved by not doing it, not as accounting costs. Thus the cost of a new transformer is what is paid for it, not its annual depreciation as recorded in the books. Costs incurred at different dates are rendered comparable by being discounted to obtain their present values, the discount rate used being taken as $7\frac{1}{2}\%$ in this paper.

The Annuity Approach

Another approach, instead of capitalizing future annual costs, decapitalizes capital costs into equivalent annuities, so using annual cost instead of present value as the common denominator. This is simple and is algebraically equivalent to the present value approach when the annual costs are constant from year to year. But in cases where they are not constant, e.g. they rise from year to year, schemes would not be correctly evaluated by comparing the sums of their annuitized capital costs plus the annual costs of the nth year, because the choice of n may affect the result of the comparison and this choice is arbitrary. A correct evaluation requires consideration of the whole time-stream of these annual costs over the life of the equipment and this in turn requires capitalization into Present Value.*

Inflation

The expectation of a continuing rise in the general price level inflates future costs relatively to present costs. This makes it profitable for any individual to borrow and acquire more assets now, but what he gains will be someone else's loss; from the national point of view the relative rise in future costs is a distortion. Thus future costs should be assessed on the assumption of a constant general price level.

Relative price changes which are expected do, however, reflect anticipated changes in the relative scarcity of different things. They thus ought to be allowed for, if they can be foreseen. If, for example, coal prices are expected to rise faster than oil prices, this expectation is certainly relevant to the choice between coal- and oil-fired generating stations.

There is one extremely important relative price change which can be anticipated with some confidence. This is a rise in the price of labour relative to the price of domestic manufactured products. This is simply the obverse of the expected growth in productivity and real wages.

Residual Values

When an investment project involves the displacement of an existing asset, that asset may be sold for scrap, in which case its scrap value

* It is, of course, true that the Present Value of all future annual costs could in turn be annuitized and added to annuitized capital cost, but this is needlessly complicated.

can be treated as a negative cost. But if the asset is not completely worn out and will probably be re-used, its negative cost is more complicated. Its written-down value in the books is irrelevant; what matters is what other costs it saves. Thus residual value must be looked at in terms of avoiding or deferring other costs.

Consider the example of a transformer with a useful life of y years installed in year 0 and which is planned to be displaced in year $y-r$ when it will have a remaining useful life of r years. Its residual value in year $y-r$ is defined as the excess of: (a) the present value of system costs as they would be then if it were not available for re-use; over: (b) the present value of system costs as they will be with it available.

Let C be the replacement cost of the transformer. Then if (a) it is *not* available for re-use in year $y-r$, a new transformer will have to be bought then and replaced every y years thereafter at a cost of C. If A is that y-year annuity whose present value is C, this has the same present value as a perpetual annuity of A commencing in year $y-r$.

If, (b), a transformer *is* available for re-use in year $y-r$, a new one will not have to be bought to replace it until year y. It will then have to be replaced every y years thereafter. Hence system costs will include a cost of C in year y and once every y years thereafter. This has the same present value as a perpetual annuity of A commencing in year y.

Residual value, being defined as the excess of (a) over (b) can now be seen to equal the present value at year $y-r$ of a constant annuity of A starting then and lasting for r years:

$$A_{y-r}+A_{y-r+1}\ldots+A_y$$

The capital part of the cost of buying and using a transformer for $y-r$ years is thus the present value of its initial capital cost less the present value of its residual value. This can be alternatively expressed in annuity terms as follows. By definition, C has the same present value as:

$$A_0+A_1+A_2\ldots+A_y$$

Hence by deducting the residual value annuity, we find the cost of using a new transformer for $y-r$ years to be:

$$A_0+A_1+A_2+\ldots A_{y-r}$$

Appendix

The Time Horizon

In principle, all costs stretching to infinity are relevant, while in practice only a limited number of years forward can be considered because uncertainty about system needs and about equipment likely to be available increase as the time horizon lengthens. But how many? Two possibilities are:

(*a*) going up to the year, if it exists, happenings after which have no effect upon the optimum choice preceding that time; i.e. the year, if it exists, when all possible strategies coincide

(*b*) picking a certain distant year, crediting Residual values in that year to all assets and neglecting happenings thereafter.

In the context of transformer replacement, consideration is likely to be focused on a limited range of standard transformer sizes, load growth beyond the capacity of the largest transformer being dealt with by putting in new substations. Rates of load growth are such that the relevant time horizon in the analysis of transformer choice in a substation will then be something like ten to fifteen years hence. Subsequent development of the load will require new substations and thus not affect the limited choice which is at issue here. In other words, whichever sequence of transformers is used, development may end up with the largest standard size of transformer, so that the final year of the different sequences differ only with regard to the remaining life (and hence the residual value) of this transformer. Where such differences are negligible, residual value can be ignored and the analysis follows alternative (*a*). Where they are not, residual value can be estimated and alternative (*b*) followed, the operation of the discount factor making the approximation permissible.

TRANSFORMER COSTS

Capital, Installation and Maintenance

The capital cost of a new transformer is simply its purchase price. The capital cost of using an old one from stock, or as a negative item, putting an old one into stock ready for use elsewhere, is a residual value whose nature has been explained above. Two complications which may not be worth bothering about in practice are that:

(*a*) since residual value is properly calculated on replacement cost, not historical cost, some reduction in it should be made to allow

111

for any inferiority in performance or operating cost of the displaced transformer as compared with a new one of the same rating whose price is used as the measure of replacement cost
(b) since residual value relates to the future use of a displaced transformer, the annuity which measures it should not be regarded as commencing with its displacement but with its subsequent use after a period of storage. Storage costs should be deducted, provided that they are not merely the result of accounting conventions but reflect a real use of extra resources.

The cost of installing a transformer or of replacing one transformer by another consists of the saving that would be made if the work were not done, i.e. the addition to costs from doing it. Thus there is reason to bring in a share of overheads only to the extent, if any, to which the costs classified as such vary with the number of transformer changes carried out.

Annual maintenance costs are usually set at a reasonable figure expressed as a percentage of capital cost.

Iron Losses

The iron loss of a transformer is independent of its load, depending only upon the characteristics of the transformer itself. Thus so long as the transformer is energized it absorbs a constant amount of power. Expressed in kW this is denoted F_e.

It is evident that the costs which are relevant are marginal costs, i.e. the cost of the increments in energy and capacity which are required to meet the iron loss. We shall here assume that generating costs consist of:

U_n = night marginal cost per kWh, applying during 2,920 hours each year

U_d = day marginal cost per kWh, applying during 5,840 hours each year

K_p = peak marginal cost per kW

all expressed in £. With a different cost structure the calculations would have to be modified appropriately.

An iron loss of F_e requires more than F_e extra kW at the generating station to the extent that the provision of F_e kW at the transformer increases losses between these two points. Let us denote such incremental transmission and distribution losses per kW by:

112

D_n at night
D_d during the day
D_p at peak

These incremental losses are functions of the shape of the night and day load curves of the system and of its load at time of peak. They can scarcely be estimated accurately, but a rough guess may be better than implicitly setting them all at zero.

The annual generation cost of the iron losses of any transformer, assuming it to be energized throughout the year, can now be seen to be:

$$F_e\left(\frac{U_n}{1-D_n}2{,}920+\frac{U_d}{1-D_d}5{,}840+\frac{K_p}{1-D_p}\right)$$

A more approximate version is obtained by averaging the night and day marginal costs into U and the three incremental loss coefficients into D, giving:

$$\frac{F_e}{1-D}(U\,8{,}760+K)$$

The capital cost of reinforcing the transmission and distribution system to provide the extra losses is the product of the iron loss of the newly installed transformer (less the iron loss of any transformer it replaces) and the capital cost, Z, per kW of network reinforcement between the generating station and the transformer:

$$ZF_e$$

Copper Losses
Copper losses are still more complicated, since they are not a constant but are for any given copper loss at full load proportional to the second power of the load in kW. A correct estimate would therefore require detailed knowledge of the transformer's phase balance and of its load curve. Since this is usually lacking, approximation has to be resorted to.

Copper losses at any time are equal to full load copper loss kW (denoted C_u) times the square of the transformer's utilization factor, i.e. to:

$$C_u\left(\frac{\text{Demand}}{\text{Rated Capacity}}\right)^2$$

so that if B_d and B_{d-1} represent the demand on the transformer at system peak this year and last year respectively and if R denotes Rated Capacity, the kW element in the generation cost of copper losses is:

$$\frac{K_p}{1-D_p} C_u \left(\frac{B_d^2}{R^2}\right)$$

for this year. The capital cost of network reinforcement necessary to supply this year's increase in losses is:

$$ZC_u \left[\frac{B_d^2}{R^2} - \frac{B_{d-1}^2}{R^2}\right]$$

if distribution peak coincides with system peak. If it does not, demand at distribution peak must be substituted for B_d and B_{d-1}. Note that this cost may be negative when a large transformer replaces a smaller one. This does not deprive the item of meaning, however, since in a growing system a reduction in demand at system peak will reduce the reinforcement expenditure which is necessary that year.

In order to approximate the average hourly copper loss over a period, the maximum hourly loss during that period is multiplied by the loss load factor of that period. Thus denoting the day hours loss load factor by L_d, the cost of the year's daytime copper losses is:

$$\frac{U_d}{1-D_d} L_d C_u \left(\frac{B_d}{R}\right)^2 5,840$$

If distribution peak and system peak do not coincide, it is the former which is relevant there instead of B_d.

Similarly, denoting the maximum demand on the transformer during night hours by B_n and the night hours loss load factor by L_n, the cost of night hours copper losses is:

$$\frac{U_n}{1-D_n} L_n C_u \left(\frac{B_n}{R}\right)^2 2,920$$

Since the number of night units is often much less than the number of day units it may be a justifiable simplification to price all units together at a weighted average of $U_n/1-D_n$ and $U_d/1-D_d$ and to take a single loss load factor. Even so, it is necessary to know not only the maximum demand during the year and (in order to estimate

114

a loss load factor) the average load factor over the whole year, but also whether the demand on the transformer at system peak B_d differs significantly from its own peak load.

Cooler Losses

In principle, the energy required by any fans or pumps associated with the transformer cooling system should also be considered, but in practice it may be legitimate to neglect this. Its inclusion would present little difficulty.

Numerical Example

Consider a sequence which involves replacing a 300 kVA transformer by a 500 KVA transformer in year seven and continued use of the latter for at least five years thereafter. The calculation shows costs only for years seven to twelve. Present value is calculated for year 0, and the assumptions are as follows:

	300 *kVA*	500 *kVA*
Cost		
Cost new	£480	£650 + £60 installation
Copper loss, C_u	4·6 kW	7·0 kW
Iron loss, F_e	0·7 kW	1·0 kW
$Z = £10$		

£454, the residual value of the 300 kVA transformer, is the present value of twenty-four years of that thirty-year annuity whose present value is £480.

The other assumptions are shown on the two tables containing the calculations. The first three columns of the second cost-up the total copper and iron losses grossed up for incremental primary system losses. It is assumed that national peak and maximum demand on the distribution system coincide but that the maximum demand on the transformer exceeds its demand at time of system peak.

TRANSFORMER INVESTMENT DECISIONS

Choice Between Different Makes

This is the problem of weighing up differences in capital costs against differences in copper and iron losses as between different transformers of the same rating. Thus the first step is to construct

| | Iron Losses | | | Demand | | | | | Copper Losses | | |
| | Day units | Night units | Demand at system peak | MD daytime | MD at night | Daytime loss-load factor | Night-time loss-load factor | | Daytime units | Night-time units | Peak kW |
Year	$F_e \times 5{,}840$	$F_e \times 2{,}920$	B_d	Q	B_n	L_d	L_n		$L_d C_u \left(\dfrac{Q}{R}\right)^2 \times 5{,}840$	$L_n C_u \left(\dfrac{B_n}{R}\right)^2 \times 2{,}920$	$C_u \left(\dfrac{B_d}{R}\right)^2$
6			222								1·4
7	5,840	2,920	240	250	100	0·6	0·3		6,132	245	1·6
8	5,840	2,920	259	271	109	0·6	0·3		7,205	291	1·9
9	5,840	2,920	280	294	119	0·6	0·3		8,480	347	2·2
10	5,840	2,920	302	319	130	0·6	0·3		9,984	415	2·6
11	5,840	2,920	327	346	141	0·6	0·3		11,746	488	3·0
12	5,840	2,920	353	376	154	0·6	0·3		13,871	582	3·5

	Copper and iron loss generation costs, £			Other costs, £				Total costs, £		
Year	Day units $\dfrac{0.6}{1-0.04}\dfrac{1}{240}$ $D_d = 0.04$	Night units $0.47\dfrac{1}{1-0.02}\dfrac{1}{240}$ $D_n = 0.02$	Peak kW $\dfrac{£9.275}{1-0.08}$ $D_p = 0.08$	Reinforcement for extra iron losses $\Delta F_e Z$	Reinforcement for copper losses $ZC_u \times \left[\dfrac{B_d^2}{R^2} - \dfrac{B_d - \frac{2}{3}}{R^2}\right]$	Net acquisition and installation	Maintenance	Cost in year	Present value of £1 at 7½%	Present value
7	31.2	6.3	26.2	3	-7.6	710-454	5.0	320.1	0.6028	193.0
8	34.0	6.4	29.2		2.7		5.0	77.3	0.5607	43.3
9	37.3	6.5	32.3		3.2		5.0	84.3	0.5216	44.0
10	41.2	6.7	36.3		3.6		5.0	92.8	0.4852	45.0
11	45.8	6.8	40.3		4.4		5.0	102.3	0.4513	46.2
12	51.3	7.0	45.4		5.0		5.0	113.7	0.4199	47.7

realistic assumptions about the load curve during each year of the projected life of the transformers. Once these assumptions are made, copper losses can be calculated for each year. Then these and the iron losses (adjusted upwards for incremental distribution losses) can be costed and incremental reinforcement costs of providing capacity to meet these losses can be calculated, given a figure for Z. The present value of the resulting stream of energy and reinforcement costs plus maintenance costs can be added to its purchase price and installation costs and the present value of its final scrap value subtracted in order to find the present value of all costs of using it in the manner postulated.

If the resulting figure is lower for transformer A than for transformer B this means that (at least so far as quantifiable costs are concerned) A is the better buy. Note that in making this comparison all costs which are the same for A as for B can be ignored in making the calculation, since they do not affect the *difference in* the present value of all costs.*

The Timing of Transformer Replacement
Suppose that it has already been decided that as the load at a particular transforming point grows, the existing transformer(s) will be replaced by a larger one (or two) of specified size and type. The problem then is to decide when this replacement should be done; it is, in other words, not a question what to do but merely one of when to do it.

Clearly the time will come when further growth of the load would overload the existing transformer, i.e. when security requirements necessitate the replacement. What is at issue here is whether it will reduce total system costs to undertake the replacement before this point is reached.

* Algebraically equivalent methods of making the comparison are: (1) Calculate the present value of the *difference* in costs in each year for the two transformers; (2) Calculate equivalent constant annuities instead of capitalizing to obtain present value; (3) Calculate the discount rate which would make the present value of the *difference* in costs in each year for the two transformers equal to zero and see whether this discount rate is greater or less than the standard discount rate used for economic appraisals.

Methods of making the comparison which are not algebraically equivalent and which therefore may (in the context of cost-minimizing investment choices in public enterprise) give a wrong answer include: (1) all kinds of pay-back period calculation; (2) expressing the difference in the average annual cost of losses as a percentage of the difference in capital cost; (3) using both capital cost and depreciation in a calculation.

Appendix

The problem can be expressed by asking whether replacement now is better than replacement a year hence. If it is, then it should be done now. If it is not, on the other hand, the decision can be postponed and the analysis repeated next year. Thus it is only necessary to investigate the *difference* in the present value of all costs involved in replacing then rather than now. If this difference is positive, replacement now is indicated.

In order to set out the argument, let it be assumed that transformers have a thirty-year life and that a discount rate of $7\frac{1}{2}\%$ is used. The annuitized cost of a transformer (denoted A in the section on residual value) is then $8\frac{1}{2}\%$ of its capital and installation cost (denoted C).

If a new transformer, costing C_n, is bought to replace an old transformer, with a replacement value of C_o, *next* year instead of this year, the extra costs involved are:

(a) $-0.085\ C_n + .085\ C_o$. The first of these terms is the present value of the saving from postponing purchase of the new transformer (and its subsequent replacement every thirty years) for one year; the second is the excess of the residual value of the old transformer available for re-use now over what it will be next year.

(b) $L_o - L_n$; the cost of the year's iron and copper losses if the old transformer is retained, less what they would be if the new one were installed this year.

(c) $M_o - M_n$; annual maintenance costs on the old transformer minus those on the new one.

(d) $R_o - R_n$; capital cost of reinforcement to supply the year's increase in copper losses on the old transformer minus the cost (or plus the saving) of reinforcement to supply the difference between the copper and iron losses on the old and new transformers.

Summing the above items, the net effect on the present value of all costs of replacing the old transformer a year hence rather than now is:

$$-0.085\ C_n + 0.085\ C_o + L_o - L_n + M_o - M_n + R_o - R_n$$

so that replacement now is indicated if:

$$L_o + M_o + R_o > 0.085\ (C_n - C_o) + L_n + M_n + R_n$$

119

In terms of year seven of the made-up numerical example given earlier this gives:

$$81 \cdot 1 + 4 \cdot 0 + 4 \cdot 3 > 19 \cdot 5 + 63 \cdot 7 + 5 \cdot 0 - 4 \cdot 6$$

i.e. $\qquad\qquad 89 \cdot 4 > 83 \cdot 6$

which indicates that replacement of the 300 kVA transformer by the 500 kVA one at the beginning of year seven was preferable to waiting till the end of that year.*

Where two transformers are operated in parallel in order to provide security, copper losses loom much smaller. It appears that the replacement of two smaller transformers by two larger ones is never then justified on cost grounds and should always wait until the growth of load necessitates it on account of security.

Replacing Obsolete Transformers

A very similar problem arises in respect of existing transformers which, although not overloaded and although not at the end of their physical life, have high copper and/or iron losses compared with new designs currently available. Thus the question is whether it is worth while to scrap them and replace them with new transformers now, instead of waiting until the end of their physical life.

Once again, the problem can be tackled by enquiring whether replacement now is better than replacement a year hence. Thus the analysis is the same as in the preceding section, with one exception. This is that the cost of postponing replacement of the old transformer is now the excess of its scrap value now over the present value of its scrap value one year hence. If S is scrap value, this excess is

$$S - \frac{S}{1 \cdot 075S} = 0.07S$$

Hence the formula in the previous section applies, with 0·07S substituted for 0·085 C_o.

Costing a Transformer Sequence

Given the anticipated load and its growth over time at a particular transforming point, the problem arises of choosing the sequence of

* 19·5 = 0·085 (710–480); 89·4 = *BST* cost of copper and iron losses on the old transformer in year seven; 4·0 = its annual maintenance cost, 4·3 = reinforcement cost of growth in its copper losses. Note that in practice it may be simpler to calculate the cost of the difference in losses and reinforcement than the difference in the costs.

transformers which will be used to meet it. Thus if the transformers which are used have ratings of R_1, R_2 ... R_5, two alternative possible sequences might be:

(a) R_1 for six years, then R_3
(b) R_2 for nine years, then R_4,

and so on.

Choice of the minimum cost sequence requires calculation of the present value of all the costs of each sequence for comparison. The information required for a simplified calculation along these lines is as follows:

1. Starting load, ultimate load and rate of growth
2. sizes and costs of the transformers to be used and the order in which they will be used
3. the limiting load or age at which each of the transformers used must be replaced
4. iron loss and copper loss at full load of each transformer
5. loss load factor, kW and kWh energy costs
6. annual maintenance costs of each transformer
7. discount rate for calculating present worth.

This involves several simplifications as compared with the discussion of principles in the earlier part of this paper. Night and day losses are not distinguished, the capital cost of network reinforcement is disregarded and maximum demand on the transformer is assumed to coincide with the maximum demand which determines the kW charge. Residual value, however, is treated in accordance with the principles outlined above.

Choosing the Optimum Sequence
If five transformer sizes are available and if no two sizes can be used together, there are sixteen possible transformer sequences for a load which starts below the capacity of the smallest and eventually grows sufficiently to require installation of the largest. If two different sizes can be used together, the number of possible sequences is even larger. There is therefore a need for a computer program to find the optimal sequence for any given load growth.

INDEX

After diversity maximum demand, 72
Amortization, 56–8
Annuity, 54, 109
Availability, 17, 50, 61, 64–6

Boiteux, M., 57

Consumer classes, 106
Consumers' surplus, 97–8, 103–4
Cost classification, 2, 15, 72, 77, 79

Depreciation, 56–8
Distribution, 68 ff.
Distribution losses, 79–81
Drèze, J., 91
Dupoux, M., 84

Economies of scale, 32, 51, 71

Fanshel, S., 10
Farrell, M. J., 87
Financial constraints, 89–91
Fixed other works costs, 15
Forecast load, 13, 21, 60
Fuel savings, 13–15

Garver, L. L., 9

Linear programming, 4, 37–8
Load duration curve, 30, 47
Load forecast, 13, 21, 60
Load increment, 44
Location of plant, 24–5, 33 ff.
Long-run marginal cost, 45 ff.

Losses, transmission, 7, 10, 22–4, 33–7, 49
Lynes, E.S., 10

Marginal conditions for optimality, 17–18, 23, 25, 29
Marginal cost curve, 14, 20, 29, 34, 36, 63
Marginal cost, long run, 45 ff. 79, 91–3, 98–9
Marginal generating cost, 2
Maximum demand tariff, 94, 100, 105
Merit order, 9, 14–15
Mix, optimal, 12 ff., 28 ff., 46–7
Mohring, H., 91

Nelson, J. R., 91
New business cost, 72, 73
Nodes, 1, 5, 31, 48

Optimization,
  hourly, 9–11
  plant mix, 12 ff., 28 ff., 46–7
  short-run, 3 ff.
  two-stage, 27

Pallister, N. D. A., 7
Peak pricing, 99–102
Programming, 4, 37–8

Rees, R., 90
Reinforcement cost, 72, 73–7
Reliability, 83–5
Reserve requirements, 64–7
Residual value, 41–2, 58, 109–11

Scott, E. C., 2
Second best, 87–9
Second-order conditions, 19–20
Short- and long-run costs, 91–3

Tariffs, 94 ff.
Technical progress, 38, 39-42, 56,
    61
Tibbarts, N., 66
Total costs, 46, 51–4
Transformers, 108 ff.

Transmission losses, 7, 10, 22–4,
    33–7, 49
Types of plant, 16

Vassall, G. S., 66
Voltage, 95

Warford, J. J., 104
Weather, 60, 64, 99
Westfield, F. M., 11
Williamson, O., 101

Milton Keynes UK
Ingram Content Group UK Ltd.
UKHW031152141024
449569UK00024B/863